Smart Nanomaterials Technology

Series Editors

Azamal Husen, Wolaita Sodo University, Wolaita, Ethiopia

Mohammad Jawaid, Department of Chemical and Petroleum Engineering, United Arab Emirates University, Al Ain, United Arab Emirates

Nanotechnology is a rapidly growing scientific field and has attracted a great interest over the last few years because of its abundant applications in different fields like biology, physics and chemistry. This science deals with the production of minute particles called nanomaterials having dimensions between 1 and 100 nm which may serve as building blocks for various physical and biological systems. On the other hand, there is the class of smart materials where the material that can stimuli by external factors and results a new kind of functional properties. The combination of these two classes forms a new class of smart nanomaterials, which produces unique functional material properties and a great opportunity to larger span of application. Smart nanomaterials have been employed by researchers to use it effectively in agricultural production, soil improvement, disease management, energy and environment, medical science, pharmaceuticals, engineering, food, animal husbandry and forestry sectors.

This book series in Smart Nanomaterials Technology aims to comprehensively cover topics in the fabrication, synthesis and application of these materials for applications in the following fields:

- Energy Systems—Renewable energy, energy storage (supercapacitors and electrochemical cells), hydrogen storage, photocatalytic water splitting for hydrogen production
- Biomedical—controlled release of drugs, treatment of various diseases, biosensors,
- Agricultural—agricultural production, soil improvement, disease management, animal feed, egg, milk and meat production/processing,
- Forestry—wood preservation, protection, disease management
- Environment—wastewater treatment, separation of hazardous contaminants from wastewater, indoor air filters.

Rakesh Kumar Bachheti · Archana Bachheti ·
Azamal Husen
Editors

Copper-Based Nanomaterial for Agricultural Practices

Springer

Editors
Rakesh Kumar Bachheti
Centre of Molecular Medicine
and Diagnostics, Saveetha Dental College
and Hospitals, Saveetha Institute
of Medical and Technical Sciences
Saveetha University
Chennai, India

Archana Bachheti
Department of Environment Science
Graphic Era (Deemed to be University)
Dehradun, Uttarakhand, India

Azamal Husen
Wolaita Sodo University
Wolaita, Ethiopia

Graphic Era (Deemed to be University)
Dehradun, Uttarakhand, India

ISSN 3004-8273 ISSN 3004-8281 (electronic)
Smart Nanomaterials Technology
ISBN 978-981-96-8609-4 ISBN 978-981-96-8610-0 (eBook)
https://doi.org/10.1007/978-981-96-8610-0

This Springer imprint is published by the registered company Springer Nature Singapore Pte Ltd.
The registered company address is: 152 Beach Road, #21-01/04 Gateway East, Singapore 189721, Singapore

Preface

The advent of nanotechnology has revolutionized numerous sectors, including agriculture, where it holds immense potential to address pressing challenges such as food security, environmental sustainability, and efficient resource utilization. Among various nanomaterials, copper-based nanomaterials (CBNMs) have emerged as promising candidates for enhancing agricultural practices due to their unique properties, such as antimicrobial activity, catalytic efficiency, and biocompatibility. This book, Copper-Based Nanomaterials for Agricultural Practices, explores the multifaceted applications of CBNMs in agriculture, their benefits, and associated challenges, offering a comprehensive understanding of their role in modern farming systems.

The book is divided into thirteen chapters, beginning with an overview of CBNMs and their relevance to agriculture (Chapter "Overview of Copper-Based Nanomaterials in Agriculture"). Subsequent chapters delve into critical topics such as the environmental impacts (Chapter "Environmental Impacts of Copper-Based Nanomaterials in Agriculture") and toxicity concerns (Chapter "Toxicity and Safety Concerns of Copper-Based Nanomaterials in Agriculture") of CBNMs, providing a balanced perspective on their safe and sustainable use. Innovative applications like pesticide detection (Chapter "Role of Copper Nanomaterials for Pesticide Detection"), crop production enhancement (Chapter "Applications of Copper-Based Nanomaterials in Crop Production"), and soil remediation (Chapter "Role of Copper-Based Nanomaterials in Soil Remediation") are discussed in detail. Chapters "Use of Copper-Based Nanomaterials for Nutrient Delivery and Uptake"–"Use of Copper-Based Nanomaterials in Postharvest Management" focus on specific roles of CBNMs in nutrient delivery, seed germination, disease management, and post-harvest management, highlighting their potential to improve agricultural productivity. Specialized topics such as their use against nematodes (Chapter "Use of Copper Nanomaterials Against Nematodes"), effects on soil microbial communities (Chapter "Effect of Copper Nanomaterials in Soil Microbial Communities"), and applications in precision agriculture and smart farming (Chapter "Copper-Based Nanomaterials for Precision Agriculture and Smart Farming") underscore the cutting-edge advancements in this field.

This book aims to serve as a valuable resource for researchers, academics, agricultural practitioners, and policymakers interested in leveraging CBNMs for sustainable agricultural development. By addressing the opportunities and challenges of integrating nanotechnology into agriculture, it aspires to contribute to the creation of innovative solutions for global agricultural and environmental sustainability. We are grateful to all contributors for readily accepting my invitation, sharing their knowledge in specialized areas of research, and readily adjusting the suggestions for improving the shape of their contributions. We hope that the book in hand becomes a valuable tool in advancing sustainable agriculture practice through nanotechnology.

Dehradun, India — Rakesh Kumar Bachheti
Dehradun, India — Archana Bachheti
Wolaita, Ethiopia/Dehradun, India — Azamal Husen

About This book

Discover the transformative potential of copper-based nanomaterials in modern agriculture with Copper-Based Nanomaterials for Agricultural Practices. This comprehensive book delves into cutting-edge applications, from enhancing crop production and nutrient delivery to controlling plant diseases and improving post-harvest management. With a strong focus on sustainability, it addresses the environmental impacts and safety concerns of nanomaterials, offering practical insights for responsible use. Bridging the gap between nanotechnology and agricultural sciences, this book provides an interdisciplinary perspective ideal for researchers, agronomists, and policymakers. Explore the role of copper nanomaterials in precision agriculture, soil remediation, and smart farming. Whether you're a scientist, practitioner, or academic, this book equips you with innovative solutions for sustainable and efficient farming. Elevate your understanding of agricultural nanotechnology and drive progress in sustainable agriculture with this essential resource.

Overview

- Covers diverse applications of copper-based nanomaterials in agriculture;
- Highlighting the role of copper-based nanomaterials in insect pest management, plant disease control, soil remediation, crop nutrition, and growth promotion;
- Discuss the toxicity concerns copper-based nanomaterials;
- Bridges nanotechnology and agriculture for innovative solutions.

Contents

About the Editors

Rakesh Kumar Bachheti graduated from Hemwati Nandan Bahuguna University, Garhwal, India, in 1996. He completed his M.Sc. in Organic Chemistry from the same university in 1998. Later, in 2001, he underwent a one-year Post Graduate Diploma in Pulp and Paper Technology from Forest Research Institute, Dehradun, India. Subsequently, he pursued his Ph.D. in Organic Chemistry from Kumaun University, Nainital, India, which he obtained in 2007. Bachheti has a total research and teaching experience of 20 Years. Before joining Addis Ababa Science and Technology University (AASTU) in Ethiopia, he worked as Dean of Project (Assistant) at Graphic Era University in Dehradun, India. It's worth noting that Graphic Era University is accredited with an 'A+' grade by the National Assessment and Accreditation Council (NAAC). Bachheti is also a Research Fellow at INTI International University Persiaran Perdana BBN, Putra Nilai, Nilai, Negeri Sembilan, Malaysia, and Adjunct Faculty at the Department of Allied Sciences, Graphic Era Hill University, Dehradun, India. Bachheti has an impressive research background and has presented papers at various international (Malaysia, Thailand, and India) and national conferences. He has also been an active member of important committees like the Internal Quality Assurance Cell (IQAC) and the Anti-ragging Committee. His research interests primarily focus on natural products for industrial applications, biofuels and bioenergy, green synthesis of nanoparticles and their applications, and pulp and paper technology. His passion for natural products is evident in all of his research endeavors. As an academic mentor, Bachheti has successfully advised 40 M.Sc. and 6 Ph.D. students to completion. Additionally, numerous undergraduates have conducted research in his laboratory under his guidance. He actively contributes to curriculum development for B.Sc./M.Sc./Ph.D. programs. Bachheti is also recognized for his editorial work, having edited 8 books and authored 150 publications on various aspects of natural product chemistry and nanotechnology. He has contributed thirty book chapters published by prestigious publishers like Springer, Elsevier, Tylor and Francis and Nova Publisher. He is listed among the Top 2% Scientists for the year 2024 by Stanford University, USA, and Elsevier.

Archana Bachheti completed her B.Sc. in 1997 and M.Sc. in 1999 from Hemwati Nandan Bahuguna Garhwal University. She earned her Ph.D. from the Forest Research Institute, Dehradun, India, in 2006. Currently, she is a Professor at Graphic Era University in Dehradun, India. With over 20 years of extensive experience, she has served in various academic roles in India and has also provided expertise internationally. Her teaching portfolio includes subjects such as ecology and environment, environmental science, freshwater ecology, disaster management, and bryophytes and pteridophytes. Professor Bachheti's research interests span a broad and interdisciplinary field of plant ecology, with a focus on ecorestoration, green chemistry—particularly the synthesis of nanomaterials—and exploring the medicinal properties of plants. Her research has involved ecological restoration of degraded lands, studying the physical and chemical properties of plant oils, and investigating plant-based nanomaterials. Throughout her career, she has been actively involved in mentoring students and researchers. Dr. Bachheti has published more than 130 research papers and edited over 8 books. She is listed among the Top 2% Scientists for the year 2024 by Stanford University, USA, and Elsevier.

Azamal Husen is served as Full Professor and Head of the Department of Biology, University of Gondar, Ethiopia; and currently working as a Foreign Delegate at Wolaita Sodo University, Wolaita, Ethiopia. He is also working as an Adjunct Professor at Graphic Era (Deemed to be University), Dehradun, Uttarakhand, India. He also worked as Full Professor at Sankalchand Patel University, Visnagar, India; and Visiting Faculty of the Forest Research Institute and the Doon College of Agriculture and Forest at Dehra Dun, India. His research and teaching experience of 25 years encompasses Nanobiotechnology, Plant Physiology, Environmental Botany, Medicinal Plants; and Clonal Propagation. He has conducted research sponsored by the World Bank, the NATP, the ICAR, the ICFRE, and the JBIC. Husen has published extensively (over 300, which includes research papers, books, book chapters etc.) and served on the Editorial Board and as reviewer of reputed journals. He has given numerous talks and presentations in many conferences and workshops. Husen has also given various orientation courses for academic staff, prepared different modules, developed B.Sc., M.Sc. and Ph.D. curriculum. Moreover, he has received a recognition award from University of Gondar, Ethiopia for a decade of service in teaching, research and community services. He is Editor-in-Chief of the *American Journal of Plant Physiology*, and a Series Editor of *Exploring Medicinal Plants* (Taylor & Francis Group, USA); *Plant Biology, Sustainability, and Climate Change* (Elsevier, USA); and *Smart Nanomaterials Technology* (Springer Nature, Singapore). He has achieved the distinguished honor of being recognized as one of the "World's Top 2% Scientists" for the years 2022, 2023, and 2024 by Stanford University, USA. Moreover, he has also worked as a special issue guest editor on various topics such as Advances in Plant Metabolomics: Insights, Applications, and Integration (Plant Physiology and Biochemistry), Exploring Nanotechnology in Medicinal Plants (*Journal of Herbal Medicine*), Integrating Nanotechnology with Plant Compounds: Enhancing Abiotic Stress Resilience and Crop Improvement via Signaling Molecules (Frontiers in *Plant Science*), etc.

Overview of Copper-Based Nanomaterials in Agriculture

Addisu Tamir Wassie, Tsige Tekle Tegie, Ibrahim Nasser Ahmed, Rakesh Kumar Bachheti, Mesfin Getachew Tadesse, Aselefech Sorsa Wotango, and Archana Bachheti

Abstract Nowadays, the global demand for food has significantly increased due to the rapid growth of the population. Furthermore, crop losses caused by diseases and pests have been exacerbated by the effects of climate change. The use of conventional agrochemicals and inappropriate transportation methods has also resulted in severe pollution of soil and aquatic environments, causing substantial environmental damage. Nanotechnology is an emerging field with significant potential for advancing agricultural development and environmental remediation. The application of nanotechnology, particularly metal-based nanomaterials, in agriculture is fundamental for improving agricultural productivity and addressing environmental challenges. Among various metal-based nanomaterials, copper-based nanoparticles hold particular importance in agriculture. These nanoparticles are anticipated to provide a viable alternative for enhancing agricultural yields and controlling plant diseases. The challenges of improving the quantity and quality of agricultural products, along with addressing environmental issues, can be tackled by enhancing plant nutrition and resistance through the use of nanotechnology and nanomaterials in crop management. The aim of this chapter is to explore the diverse applications of copper-based nanomaterials in agriculture, focusing on their properties, synthesis methods, environmental impact, and safety. Finally, this chapter highlights contemporary challenges associated with the use of copper-based nanomaterials and offers perspectives on future research directions.

A. T. Wassie · T. T. Tegie · I. N. Ahmed · M. G. Tadesse · A. S. Wotango
Department of Industrial Chemistry, College of Natural and Applied Sciences, Addis Ababa Science and Technology University, Addis Ababa, Ethiopia

A. T. Wassie
Department of Chemistry, College of Natural Sciences, Wollo University, Dessie, Ethiopia

R. K. Bachheti (✉)
Centre of Molecular Medicine and Diagnostics (COMMD), Saveetha Dental College and Hospitals, Saveetha Institute of Medical and Technical Sciences, Saveetha University, Chennai, Tamil Nadu, India
e-mail: rkbachheti@gmail.com

A. Bachheti
Department of Environment Science, Graphic Era University, Dehradun, Uttarakhand, India

R. K. Bachheti et al. (eds.), *Copper-Based Nanomaterial for Agricultural Practices*, Smart Nanomaterials Technology, https://doi.org/10.1007/978-981-96-8610-0_1

Keywords Nanotechnology · Cu-based nanomaterials · Agriculture · Properties · Synthesis · Applications · Toxicity

1 Introduction

Nanotechnology is the study of materials with a size ranging from 1–100 nm. It is used in many disciplines like agriculture, paint manufacture, medicine and textile industries, chemistry, physics, engineering, and biology. The esteemed Nobel Laureate Richard Feynman first developed the nanotechnology field in his well-known presentation at the California Institute of Technology on December 29, 1959. He used a top-down method to describe the design of nanomaterials [63]. Nanomaterials are important because of their unique physicochemical properties, and in nanotechnology, they can entirely revolutionize the environment, food, and agricultural sciences [14, 38–40, 42, 75, 89]. Nowadays, these emerging technologies are pervading several scientific disciplines that have possessed the advancement of nanomaterials utilization within a few decades [31]. In the last few years, applying nanotechnology to sustainable agriculture is becoming more popular. This results from many applications it can serve in agriculture, such as establishing nanoscale tools and materials to increase yields and food security, improving water and fertilizer use efficiency via delivery systems, and reducing environmental pollution. Thus, it can be said that in the agricultural area, this promising technology is highly supportive of the formation of nanoscale agrochemical delivery systems that comprise insecticides, fertilizers, and growth regulators. The nanoencapsulation scale pie technologies have enhanced penetration and distribution, elevated effectiveness, and little environmental impact related to traditional managing methods [74]. Nano-based agrochemicals can be delivered directly to the intended location in the plant by encasing them in nanoscale particles, which minimizes waste and off-target impacts. Based on these advantages nowadays, nanotechnology has given attention to advancing the growth of harmless pesticides with nano-formulations to increase crop yield and enhance soil fertility. For instance, as an alternative to traditional $CuSO_4$, copper (Cu)-based inorganic nanopesticides, such as Kocide®3000, have been made accessible in the United States of America market and are being utilized in organic farming [61].

The use of nanotechnology in agriculture and food systems presents enormous potential for raising people's quality of life. Figure 1 illustrates the growing central role that nanomaterials are playing in the agriculture industries. A range of nanomaterials is suggested in agriculture to help minimize nutrient losses, optimize water management and nutrients, and help lower the usage of agrochemicals by using smart delivery systems. Furthermore, agricultural products or by-products can be used to create bio-nanocomposites. However, in contrast to other industrial sectors, the potential benefits of nanotechnology in agriculture are currently minimal and have yet to be widely exploited. Much research should be done in agricultural nanotechnology to find novel approaches for tracking, evaluating, and managing agricultural

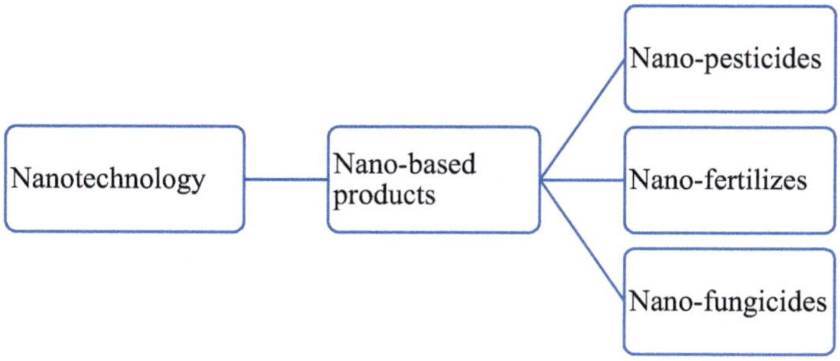

Fig. 1 Nanotechnology for the development of agrochemicals

practices that address sustainability, higher productivity, disease control, and crop protection [12, 13, 25, 37].

Increasing global food security, lowering the environmental impact of agriculture, and reducing malnutrition are severe issues in enhancing our food systems. Sustainable farming practices and crop development could benefit significantly from applying agricultural nanotechnology. Innovative active ingredients, like nano-fertilizers and micronutrients, agrochemical-encapsulated nanocarrier systems, and the delivery of nanopesticides, such as green pesticides and biopesticides, could help the agri-food sector. Because they are dispersed and leached, nonstandard agrochemical use harms the environment and endangers aquatic and terrestrial life. The goal of agrochemicals' controlled release of active ingredients is to reduce overuse by limiting the quantity and frequency of applications in the field. It is important to assess how well Cu-based nanoparticles work compared to conventional agrochemicals to tackle agriculture issues [2].

Among metal-based nanomaterials, a broad range of Cu-based nanomaterials are widely employed in agriculture. This is because, among other advantages, they are expected to present a viable substitute strategy for improving agricultural production and controlling crop illnesses [78]. Lowering the particle size to the nano dimension makes these compounds more effective since they have a more extensive surface area and can perform better as biocidal agents. The most significant Cu-based nanostructures are cupric oxide (Cupric oxide), copper sulfide (CuS), and cuprous oxide (Cu_2O) nanoparticles, which have become more and more relevant commercially because of their lower cost, simplicity of synthesis using environmentally friendly methods, customizable size and shape, and variety of uses in the agriculture sector. In the movement, Cu-based monogenic agrochemicals can promote crops and combat agricultural ailments, providing a more maintainable substitute to traditional agrochemicals. In terms of protecting plants from various pests, weeds, and plant diseases brought on by a wide range of fungal, bacterial, and viral plant pathogens, Cu-based nanomaterials have become increasingly important in agriculture.

Additionally, the Cu-based nanomaterials are used as a nematicide, insecticide, and larvicide against a variety of mosquito vectors and insects as well as plant insect pests such as spider mites, mealybugs, thrips, and leaf worms [44]. The necessity of the current study is that even if many types of research are done on the application of Cu-based nanomaterials, there is a deficiency of an organized and inclusive overview of Cu-based nanomaterials of potential applications in the agricultural field, their unique properties, synthesis methods and their toxicity in plant and soil. We also provide organized information based on our overview of the findings of the latest experimental research, along with a review of pertinent Cu-based nanomaterials literature. Finally, we summarized the benefits and challenges of using Cu-based nanomaterials in agriculture, and ideas are made for future research directions.

2 Properties of Copper-Based Nanomaterials

The presence of auspicious and unique properties is crucial to using Cu-based nanomaterials in various technological applications. Their versatility comes from the fact that they have many different properties that can be used in an array of industrial and agricultural sectors. Cu-based nanomaterials exhibit improved electrical conductivity as compared to their bulk forms. Because of their high surface area properties, which facilitate effective electron transmission, they are valuable ingredients in conductive inks for flexible electronic applications. In addition, they are used as effective catalysts in the oxidation and hydrogenation reaction and for the synthesis of fine chemicals. Mainly due to its capability to exist in variable oxidation states, copper can be used in versatile redox reactions in catalytic processes. Copper nanoparticles also have high thermal conductivity properties, which makes it essential to enhance thermal interface materials and effective heat management in electronics.

Furthermore, their optical characteristics may be tuned, which makes them appropriate for a wide range of photonic and optoelectronic uses, from light-emitting devices to sensors [57]. Cu-based nanomaterials possess good antibacterial and antifungal activity against various pathogenic fungi and can also effectively control pests in agriculture [43]. In addition to the above-listed properties, Cu-based nanoparticles' unique chemical and physical properties are discussed in the next section.

2.1 Chemical Properties of Cu-Based Nanomaterials

The common key chemical properties of Cu-based nanomaterials which makes them unique and valuable in numerous applications in industries and agricultural area are: the nature of antimicrobial properties (have ability to disrupt microbial cell membranes and can affect their metabolic processes by releasing copper ions and produce reactive oxygen species) [23], catalytic activity (because of presence of the high surface area and active sites on the surface of it) [6], surface chemistry (ability to improved easily its surface through functionalization with different types of ligands), tunable properties (enhance its activities through varying their shape or composition or size) [87], pesticidal properties (ability to disrupting physiological processes of insect like pests which causing to mortality), and slow-release of fertilizers which is assured the constant supply of nutrients to floras over time to reduce the frequency of fertilizer addition [93].

2.2 *Physical Properties of Cu-Based Nanomaterials*

The agriculture business has been into nanoparticles lately for their special physical traits, such as being very small but having a lot of surface area and electrically charged. Because of these properties are ideal for products like nano fertilizers, antimicrobial agents, and antifungal treatments. Their physical attributes make Cu-based nanomaterials useful as agrochemical tools. Additionally, exploiting the physicochemical features of Cu-based nanomaterials can improve their crop protection efficiency while changing plant health and soil conditions. Several important physical characteristics make Cu-based nanomaterial highly potential for application in agriculture: size effect at the nanoscale; particle morphology such as multi-morphological structures and stabilization effects; small-sized particles with uniform size distribution; homogeneous oxidative state with high dispersive quality nature leading to appreciable mechanical strength [68]. Almost all Cu-based nanoparticles have a particle size between 1 and 100 nm, which gives them a large surface area to volume ratio. This is critical in enhancing its performance in low quantities compared to bulk material and improving its reactivity and interaction with biological systems. As such, they have wider surface areas that make them interact more with plant pathogens and tissues as well as soil particles, thus increasing their efficiency in fertilizers, antimicrobial agents, and soil amendments [29]. Furthermore, the shape of Cu-based nanomaterials like cylinders, squares, threads, or circles can also influence their physical properties, including dispersibility, stability, and interaction with biological systems [29]. The different shapes can vary the degree of contact with plant surfaces and microbial cells, affecting Cu-based nanomaterials' antimicrobial efficacy and plant uptake [91]. Cu-based nanomaterials can form steady colloidal dispersions in aqueous solutions, which is very important to give uniform spray and treatment coverage on plant surfaces and soil in agricultural fields [10]. Cu-based nanomaterials play a significant role in monitoring environmental conditions and detecting plant pathogens by developing optical sensors. The optical properties can make a visible difference in colors as a sign of the nanoparticles' existence and distribution. This is one way to show if they are within their bounds [84]. Other physical traits relating to Cu-based nanomaterials are their high strength, meaning they can last for long in the soil and on the plant surfaces, being useful as a good source of nutrients in agriculture besides being antimicrobial agents [18].

3 Synthesis Methods of Copper-Based Nanomaterials

The synthesis method of nanomaterials is crucial in progressing nanotechnology and its numerous applications. Cu-based nanomaterials may be synthesized by conventional (physical and chemical) and biological methods. The properties of nanoparticles are influenced by the types of synthesis methods [1, 22, 36]. The Cu-based nanomaterials synthesis requires copper salt to produce copper ions, reducing agents for electron production and nanoparticle aggregation [1, 64].

The physical synthesis method of Cu-based nanomaterials utilizes electric current as the source of electrons to convert copper into its elemental form within the size range of 1–100 nm [36, 47]. The various physical synthesis methods used for Cu-based nanomaterials synthesis are chemical vapor deposition, electrodeposition, sputtering, ball milling, laser ablation, pulsed wire discharge, pulsed laser deposition, spark discharge, and arc discharge [22, 27, 45]. Due to its pure and uncontaminated nanoparticle production ability, the laser ablation technique is unique from other traditional physical synthesis methods [33]. Metal sputtering, annealing, and atomization have been reported as suitable protocols for synthesizing Cu-based nanoparticles [13, 27]. The physical synthesis methods Cu-based nanoparticles were superior to chemical synthesis methods due to the uniform distribution of nanoparticles and the lack of reagent contamination [33, 57, 82].

The chemical synthesis method is another Cu-based nanomaterials synthesis method, which can be classified into two main categories: the traditional and green chemical synthesis methods. The green chemical synthesis method uses naturally existing reducing agents, including citric acid and ascorbic acid. In contrast, the traditional synthesis method of Cu-based nanomaterials potentially uses dangerous reducing agents such as hypophosphite and sodium borohydride [86]. This synthesis method also employs electrochemical and photochemical reactions, which need high temperature, pressure, and energy [22, 50]. The chemical synthesis method has serious disadvantages like toxicity, environmental pollution, and difficulty in the process [50, 59]. Nevertheless, nanomaterials' physical and chemical synthesis methods have been widely applied due to their ability to produce monodispersed nanoparticles with controllability. However, they have many limitations that reduce their acceptance. The commonly known limitations are using expensive and hazardous chemicals, demanding extremely high pressure and temperature, consuming long reaction times during synthesis, and absorbing harmful side products on the outer layer of nanoparticles [79]. Therefore, these and any other limitations of conventional synthesis methods prompt researchers to search for a sustainable substitute method known as the green synthesis method.

Nowadays, green synthesis methods have been widely utilized to synthesize numerous Cu-based nanomaterials [59]. In recent years, the use of green (biological) synthesis methods to produce nanomaterials has increased, and it is currently the most popular synthesis approach in contemporary material sciences [53]. The biological synthesis method is the most suitable due to its low cost, cleanliness, sustainability, eco-friendliness, and low energy demand. Produced nanoparticles have excellent properties, such as being more stable and biocompatible, compared to conventional synthesized products. The final goal of the biological synthesis method is to scale up nanoparticle activity while reducing their negative impact on the environment and any living things' health [50, 88]. Green synthesis methods encourage the natural abilities of numerous biological entities, such as biomolecules, plants, and microorganisms. The drawbacks of physical and chemical nanomaterial synthesis methods, like the use of poisonous and costly chemicals, the high-energy necessities of production, and so on, are solved by this synthesis method [41]. Most research has been done to discover innovative green resources to investigate nanoparticles and their potential outcomes [59]. Green synthesis resources for manufacturing Cu-based nanoparticles include plant parts, and various microorganisms such as bacteria, yeast, fungi, and viruses [5, 51]. Mostly, the biological synthesis methods used plant extracts, microorganisms, and enzymes to provide electrons for the reduction of Cu salt and the steadiness of Cu-based nanoparticles [17].

Medicinal plants that are particularly highly rich in various phytochemicals (phenols, quinines, alkaloids, flavonoids, tannins, and terpenoids) are known to mediate nanoparticle synthesis and are very important in reducing copper salts. They are the best optional and positive intermediate to synthesize massive nanoparticles [81]. Sultana and colleagues' (2023) study showed that the biological synthesis method is the simplest method of Cu-based nanoparticle synthesis. This is they used nontoxic solvents like water for extracting the plant materials, and then the extracted plant materials were mixed with a solution of copper salt and they allowed the mixture to react until the reaction took place to produce Cu-based nanoparticles [80]. Plant extracts can influence the morphology, shape, and size of nanoparticles. For instance, Bagherzadeh and colleagues (2022) study showed that the *Rosmarinus officinalis* leaf extract served as both a stabilizing and reducing agent during capping nanoparticles with ketone and aldehyde functional groups to control the growth, aggregation, and reduction processes. According to their research, adding additional phytoconstituents results in smaller generated nanoparticles, which are influenced by the plant extract content [15]. In other ways, the plant extract-centered synthesis of nanomaterial is easier than microbial-based synthesis, even if the microbial method is more economical and less susceptible to variations in plant growth stages and seasons [80]. The general green synthesis method of nanoparticles using plant materials is depicted in (Fig. 2). Microbial-assisted synthesis of Cu-based nanomaterial using fungi, bacteria, and algae is also a biological synthesis method used for large-scale products. This method is also sustainable and environmentally benign since it is carried out in either extracellularly or intracellularly temperate conditions [52]. This synthesis method of nanoparticles uses bacteria and fungi materials with the copper salt solution, as depicted in (Fig. 3).

Enzyme-mediated synthesis of nanoparticles is another common biological synthesis method that prefers mild conditions to synthesize the Cu-based nanomaterial. They can show excellent particle size and high specificity. The process of integrating an enzyme onto a nanostructure is contingent upon various elements, such as the immobilization technique, the nature of the nanomaterial, and the operational and environmental circumstances. Enzyme-integrated nanoparticles have confirmed

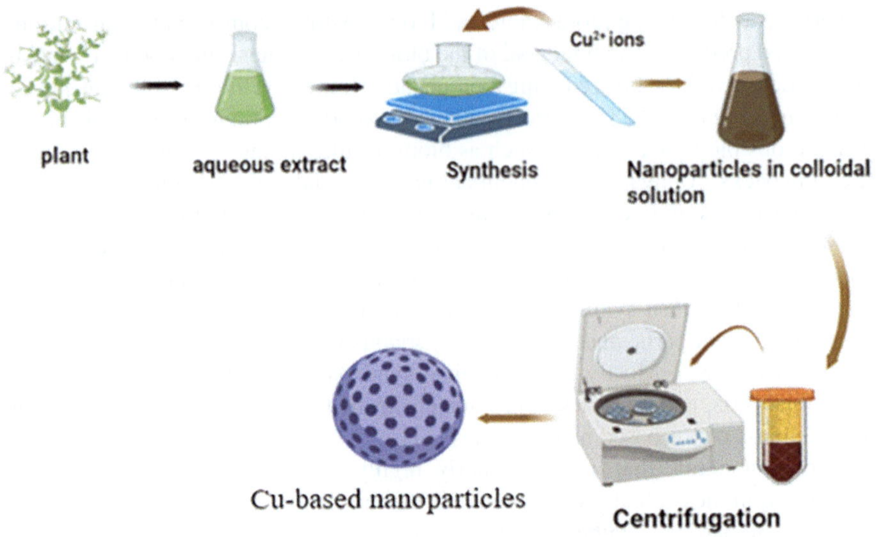

Fig. 2 Schematic diagram of Cu-based nanoparticles green synthesis using plant material [60]

encouraging outcomes in a wide range of various applications, including medication therapy, wastewater remediation, biosensors, and biocatalysts. This is because of their distinct capability to show physical and chemical characteristics, including constant catalytic activity under a variety of operating conditions, a high surface area, sensitivity, recovery or reusability rates, and response scale [8]. Carbohydrates (polysaccharides), peptides, and proteins are other common biomolecules used as mediators to stabilize and reduce copper ions in synthesizing Cu-based nanomaterials with a copper salt solution. They can produce unique nanoparticles with specific applications by providing additional functional groups [24].

4 Applications of Copper-Based Nanomaterials in Agriculture

4.1 *Nutrient Delivery and Enhancement*

Even though micronutrients are needed in comparatively small amounts for crop growth, they are very momentous for various plant metabolic reactions such as photosynthesis, nitrogen fixation, amino acid production, and cation neutralization. However, insufficient levels of micronutrients in plants cause growth reduction, leaf deformation, chlorosis, yield decrease, and lower quality of plant products. Nowadays, micronutrient fertilizers are essential because intensive agricultural practices have depleted the natural soil micronutrient pools for the past few decades. Adding

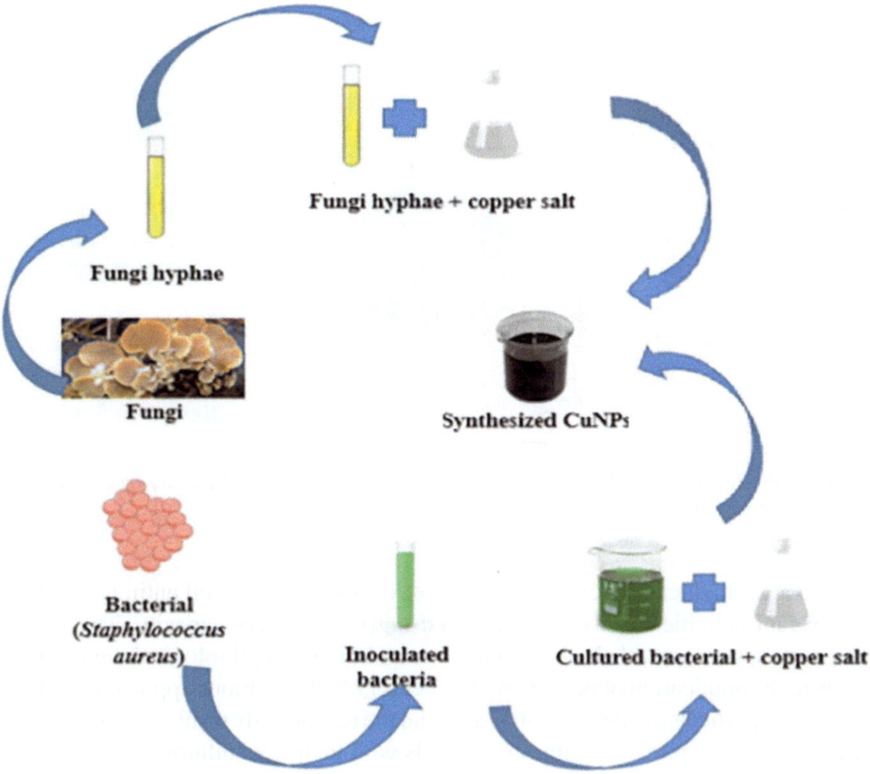

Fig. 3 Schematic diagram of Cu-based nanomaterial synthesis using bacteria and fungi [6]

micronutrients at lower rates to nanoparticle fertilizers (nitrogen, phosphorus, and potassium) was one of the earliest strategies for giving plants the nutrients they needed [56]. Therefore, various mineral elements are required for the plant nutrients in the form of micro and macronutrients responsible for macromolecules' structural components. Plant-required macronutrients with a concentration of more than 0.1% are calcium (Ca), magnesium (Mg), sulfur (S), potassium (K), nitrogen (N), and phosphorus (P) whereas micronutrients (<0.01% concentration) includes Cu, iron (Fe), zinc (Zn), boron (B), nickel (Ni), molybdenum (Mo), manganese (Mn), and chlorine (Cl) [66]. As we mentioned above, Cu is an essential micronutrient for plant growth. It is comparatively cheaper and universally available. Therefore, using Cu-based nanoparticle fertilizers in various agricultural applications is cost-effective. Furthermore, because of their small size, they are easier to absorb, penetrate, and translocate inside plant cells to enhance plant growth. Copper nanoparticles also improve the control of microorganisms and insect pests [66]. According to research by Sudhir and colleagues (2015), Cu-based nanoparticles are the most suitable materials for making nanoagrochemicals, and they have shown remarkable applications in modern agriculture. This is attributed to their unique properties such as large

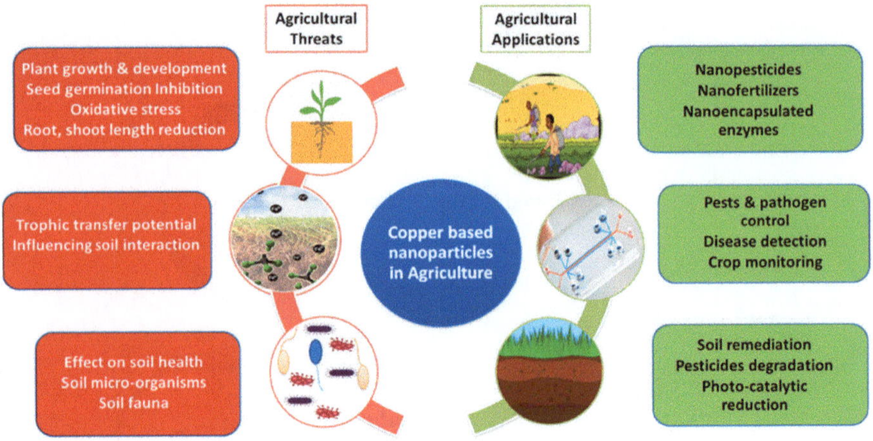

Fig. 4 Multi-directional application of Cu-based nanoparticles in agriculture and various predicted agricultural treats [15]

surface area, high catalytic activity, low cost of production, good antimicrobial and antioxidant activities as well as enhanced nutrient delivery capacity. The type of nanoparticles present, their shape and size, and the nearby biological environment regulate the application of nano-agrochemicals [73]. The various applications of Cu-based nanoparticles are depicted in Fig. 4, and the recent study results of various types of Cu-based nanoparticle synthesis methods with their agricultural applications are shown in Table 1.

Ekanayake and Godakumbura [26] synthesized a low-cost, double-functionalized fertilizer for enhancing the soil structure by retaining H_2O and supplying micronutrients simultaneously. This synthesis was achieved using a nano nutrient particle source of metal oxide nanoparticles covalently bonded with alginic acid.

The ZnO and Cupric oxide nanoparticles were metal oxides used to form micronutrient nanofertilizers. This is because more research suggests that these metal oxides have an advantage in plant growth by nourishing young plants and seedlings, influencing the flowering period, and boosting crop yield. Given all of these factors, they concluded that the dual purpose of the planned synthetic nano fertilizer is that the alginate hydrogel functions as a soil conditioner by retaining soil moisture, which improves plant roots' access to nutrients, and the solid ZnO and Cupric oxide nanoparticles act as nano nutrients for the nourishment of young plants and seedlings. Plant roots can also absorb metal oxide nanoparticles easily and then transfer to the shoots when the plants receive the nanofertilizer. Mainly, roots absorb $Cu(aq)^{2+}$ and $Zn(aq)^{2+}$ cations, which are crosslinked with alginate hydrogels and translocated to different parts of the plant to convey metabolic reactions at different times [26].

Table 1 Synthesis methods and agricultural applications of various types of Cu-based nanomaterials

S/No.	Cu-based nanomaterials	Particle size (nm)	Synthesis method	Agricultural applications	Key references
1	Cu/Zn bimetallic nanoparticle	6	Chemical	To enhance growth and germination of chickpeas	[54]
2	Cu-chitosan nanoparticle	Not reported	Chemical method	For plant growth promoting and enhancing mechanism of plant defense	[20]
3	CuO-nanoparticles	28	Biosynthesized	As a nano-fertilizer, antimicrobial agent, and plant growth stimulant	[77]
4	Cu_2O, Cupric oxide, and Cu/Cu_2O nanoparticles	11–14	Hydrothermal	As antifungal activity against *Phytophthora infestans*	[31]
5	Cu-loaded silica nano gel (CuSiNG)	4–8	Acid-catalyzed sol–gel method	To improve antimicrobial efficiency in *X. alfalfa and E. coli*	[92]
6	Chitosan/polyacrylic acid/Cu nanocomposites	6.7	Polymerization	To improve the growth and yield of onion bulbs	[4]
7	CuO/polyelectrolyte complexes nanoparticle	300	Encapsulation	Regulate Cu fertilization without dispersion in the soil; seed germination promoter	[49]
8	CuO nanoparticles-embedded hydrogels	Not reported	By loading Cupric oxide nanoparticles in hydrogels	Enhance disease resistance of plants, increase nutrient uptake, and improve plant growth	[72]
9	Cu-chitosan nanocomposite hydrogels	Not reported	In situ reaction	For against aflatoxigenic from dairy cattle feed *(aspergillus flavus)*	[3]
10	Cu-chitosan nanoparticle	Not reported	Ionic gelation (chitosan-sodium tripolyphosphate)	As antifungal agents: *Fusarium wilt* of tomato)	[69]

(continued)

Table 1 (continued)

S/ No.	Cu-based nanomaterials	Particle size (nm)	Synthesis method	Agricultural applications	Key references
11	Nano-ZnO–Cupric oxide alginate	Not reported	Chemical	Nanofertilazer: to provide micronutrients to crop plants while nourishing the soil to an organic state	[26]
12	Cu/Zn nanoparticle	59–119	Biosynthesis	High efficacy against *Phenacoccus solenopsis* (insect)	[55]
13	Cu/Cu_2O nanoparticle	56	Aqueous-phase synthesis	Better antifungal activity than Cupravit®Hidro (commercial agricultural fungicides)	[34]
14	CuO nanoparticle	40–80	Biosynthesis	ATP production interferes by preventing biofilm formation in bacterial cells	[19]
15	CuS nanoparticle	36.1 ± 6.4	Chemical	To reduce disease occurrence in rice, both the seed and plant stages	[71]

4.2 Cu-Based Nanoparticles as Crop Protection

Antibacterial, Antifungal, and Insecticidal Activities

Crops can be affected by different microbes, fungi, insects, and so on. So, modern agricultural technologies use Cu-based nanoparticles to prevent or make crops free from such diseases. Figure 5 shows the physiological effects of Cu nanoparticles on pest control, namely fungi, insects, bacteria, and nematodes. Improved comprehension of the principles underlying Cu-based nanoparticles' antibacterial action and plant nutritional regulation would significantly improve the ways of managing agricultural diseases with nanotechnology [72]. Cu-based nanoparticles are commonly known for displaying antimicrobial characteristics and strengthening their agricultural efficiency for defensive crops against plant diseases and pathogens. The small size and ion-releasing ability are other basic concerns that show a better antimicrobial activity with a broad spectrum for defending crops against pests due to their various actions against numerous microorganisms. They are easily absorbed into plants due to their small size, enhancing their biological efficacy and avoiding the risks of overuse and runoff. Lastly, this might lessen the general adverse effects of farming on the agroecosystem, especially the soil system [16].

Cu-based nanoparticles are also used as fungicides, which are crucial in preventing disease in a wide range of plant species because they produce reactive radicals (hydroxyl) that can harm proteins, deoxyribonucleic acid, lipids, and other biomolecules [66]. Many research teams have synthesized Cu-based nanoparticles and examined how they work against pathogens that cause plant diseases, fungi, and bacteria. For example, nanocomposites based on Cu-chitosan (Cu@Chit NCs) have been synthesized using the metal-vapor synthesis method to test the antifungal activities of chitosan-impregnated Cu nanoparticles. The inhibition of growth by the Cu@ChitNCs was observed against *Rhizoctonia solani* AG-4 and *Sclerotium rolfsii*, two plant-pathogenic fungi. They destroy hyphae and induce cytoplasmic coagulation by preventing fungal multiplication.

On the other hand, the ionic gelation technique using chitosan and tripolyphosphate ion cross-linking was employed in synthesizing copper-chitosan nanoparticles. Using *Alternaria solani* and *Fusarium oxysporum* fungus species as examples, Saharan and colleagues' (2015) study sought to determine how these composites affected tomato plants' growth rates. Finally, the reduced severity of disease symptoms in fungal spore germination and mycelia development caused by composites were regarded as inhibition ability [3, 69]. Numerous studies have demonstrated that cu-based nanomaterials are also used in agriculture to control insects. Copper (II)

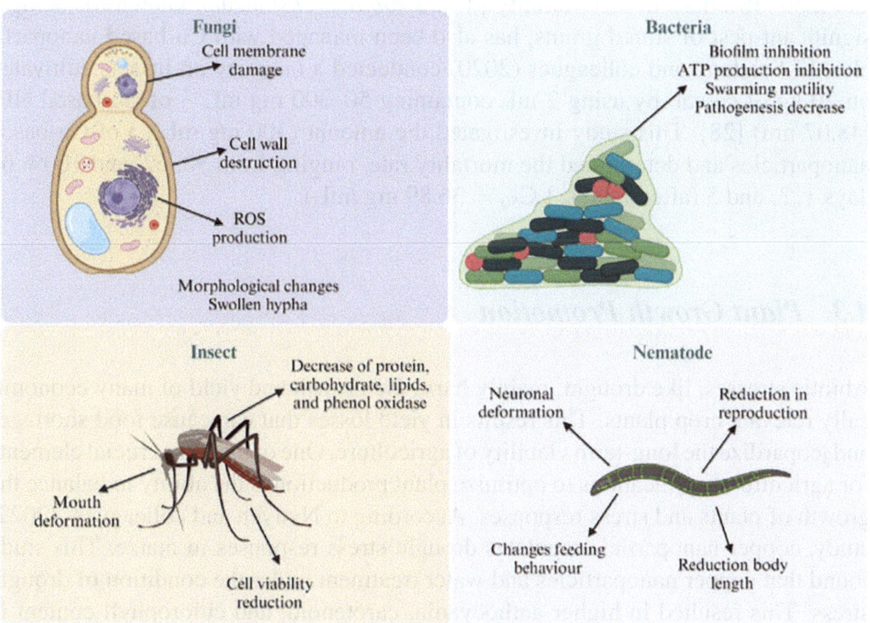

Fig. 5 Copper nanoparticles' physiological effects on fungi, insects, bacteria, and nematodes in pest control [82]

oxide nanoparticles that can be fed to insect pests have been emerging as one of the methods due to their cost-effectiveness and environmentally friendly technology [66]. One strategy of this kind is changing nanoscale properties such as dissolving rate so that Cu-based nano pesticides can find a better way for efficient delivery, higher crop yields, and reduced environmental negative impacts [72]. In another research conducted by Atwa and co-authors (2017), larval mortality was induced by copper oxide nanoparticles that were ingested artificially in *Spodoptera littoralis*. In particular, 40% of larval mortality was recorded on day one, but 100% of larval mortality was seen on day four when 50–300 mg doses were added to 150 mL of artificial feed. Another study investigated copper oxide and copper oxide doped with silver nanoparticles (nanosheet) cumulative lethal effect by feeding. Even though both copper oxide and copper oxide doped with silver nanoparticles show cumulative larval mortality, they had no instantaneous effect. However, when they were applied in high concentration for a very prolonged time, the number of *S. littoralis* larvae was reduced significantly. The maximum insect mortality rate was attained with 300 mg of nanosheets; among copper oxide-treated 3rd and 5th instars, it was 97.6% and 100%, whereas, for copper oxide: Ag-treated instars, it was 57.6% and 48.2%, respectively. Finally, they concluded that copper oxide was more potent than copper oxide doped with silver at all concentrations and with a similar exposure time [11]. Furthermore, Leon Jimenez and colleagues' (2019) study showed that Cu-based nanoparticles significantly affect *Phenacoccus solenopsis* Tinsley. This was indicated by the data with toxicity after 96 h of treatment, a 40% mortality rate, and a 30–38% loss in cell viability [48]. *Tribolium castaneum*, known for being a significant pest of stored grains, has also been managed with Cu-based nanoparticles. El-Saadony and colleagues (2020) conducted a bioassay on insects cultivated on 20 kg of cereal, by using 2 mL containing 50–300 mg mL^{-1} of Cu-based NPs (48.07 nm) [28]. This study investigated the amount (300 mg mL^{-1}) of Cu-based nanoparticles and determined the mortality rate, ranging from 30, 67, and 100% on days 1, 3, and 5 (after 5 days, $LC_{50} = 36.89$ mg /mL).

4.3 Plant Growth Promotion

Abiotic stresses, like drought, mainly harm the growth and yield of many economically relevant crop plants. This results in yield losses that can cause food shortages and jeopardize the long-term viability of agriculture. One of the most crucial elements for agricultural applications to optimize plant production is the ability to balance the growth of plants and stress responses. According to Nguyen and colleagues' (2022) study, copper nanoparticle regulates drought stress responses in maize. This study found that copper nanoparticles and water treatment under the condition of drought stress. This resulted in higher anthocyanin, carotenoid, and chlorophyll content in copper nanoparticle-treated plants than in water-treated plants. Furthermore, histochemical tests using 3,3-diaminobenzidine (DAB) and nitroblue tetrazolium (NBT) revealed a decrease in reactive oxygen species (ROS) content on priming plants.

This was attributed to an increase in ROS-scavenging enzyme activities throughout drought seasons. Also, analyzing their relative yield data showed that when subjected to droughts, copper nanoparticle use led to an increased number of seeds per plant and grain production [83]. Shafiq and colleagues (2024) also examined the effect of copper oxide nanoparticles on the growth and metabolic traits of two maize hybrids FH-1046 and YH-5427 under normal circumstances as well as saline stress. In this respect, the photosynthetic parameters like stomatal conductance, transpiration rate, and photosynthetic rate were significantly decreased by salt stress but Na concentration increased in shoot and root. These researchers investigated the improved photosynthesis and growth qualities along with the K, N, Mg, P, and Ca amount in the roots and shoots when they foliar spray with copper oxide nanoparticles [70].

5 Toxicity of Cu-Based Nanoparticles in Plants and Soil

As we discussed above section, Cu-based nanoparticles have a wide range of agricultural applications, including antibacterial, antifungal, and insecticidal activities for plant growth promotion, nutrient delivery and enhancement, and so on. Although Cu-based nanoparticles have so many distinct applications in different agricultural fields, they have also participated with some toxicity belongings (Fig. 6). Their persistent property and long-term use in agriculture may expose the agricultural soil system, particularly, to various unknown risks to the agroecosystem. This covers the effects on soil microorganisms' number, variety, community, and other soil processes such as microbially mediated soil cycles. Concerns are also raised by their effects on plants, rhizospheric interactions, consequences for the food chain, and the possibility of trophic transfer. In addition, their accumulation in the soil and their interactions with other elements of the soil, such as contaminants, may affect the general health and quality of the soil system [16].

Even if Cu-based nanoparticles can promote plant growth and inhibit phytopathogens, they negatively affect plants, particularly edible crops, growth, development, and physiological functions when they are found elevated levels [35]. Many studies showed that the effects and phytotoxicity of Cu-based nanoparticles differ depending on the plant type. Dimkpa and colleagues (2015) investigated the bioaccumulation factor of Cu-based nanoparticles in plants and suggested that root morphology influences bioavailability. According to this study, different levels of Cu-based nanoparticle translocation in alfalfa and lettuce and the toxicity of cupric oxide on carrots are determined by their unique taproot anatomy and a root vegetable [21]. Plant root biomolecules may facilitate the availability of nanoparticles by interacting with enzymes and chelating agents. Plenty of studies have demonstrated that roots accumulate the highest concentration of Cu-based nanoparticles [7, 62]. Root morphology is thought to influence rhizospheric interactions, which cause differences in plant species. The application mode and exposure medium, like soil, can influence Cu-based nanoparticle uptake, translocation, phytotoxicity, and plant accumulation [82].

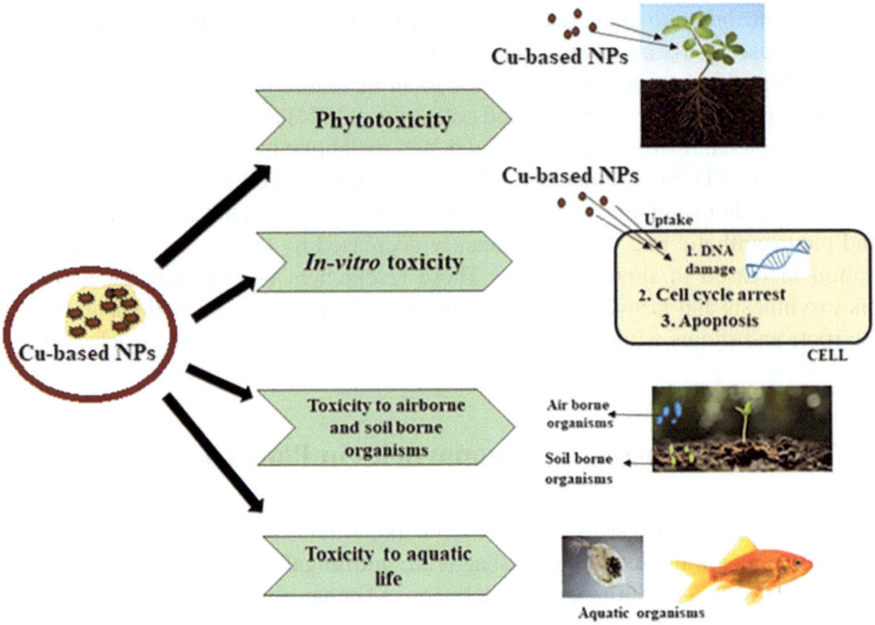

Fig. 6 Toxicity of Cu-based nanoparticles (Cu-based NPs)

As many studies have shown, the cause of the dramatic decrease in plant root length and chlorophyll amount is the transfer of Cu-based nanoparticles to plant stems and leaves. The accumulation of higher concentrations of Cu-based nanoparticles exhibited at the root zone had a significant impact on the rhizospheric microenvironment in addition to restricted Cu translocation in plant tissues. The plant tissues' penetrating power of Cu-based nanoparticles may also influence the yield of leafy crops. As Xiong and colleagues' (2017) study indicated, a high uptake of Cu-based nanoparticles has been observed in lettuce and cabbage. This high accumulation of nanoparticles lowers plant weight, photosynthesis, and water content [90]. Cu-based nanoparticles can inhibit the growth of *Glycine max* L physiological processes, including altering genes associated with oxidative stress and ROS production. Microparticles based on copper inhibit the germination of *Coriandrum sativum* seeds. Similarly, they prevent *Oryza sativa* from germinating and further lower stomatal conductance, transpiration rate, and photosynthetic parameters [85].

Furthermore, dissolved free metal ion species, which are toxic and readily bioavailable, are typically linked to the toxicity of conventionally high soil concentrations of Cu-based nanoparticles [7]. On the other hand, Rajput and colleagues (2020) suggest that Cu-based nanoparticles may have distinct toxicity characteristics that involve a different mode of toxicity, which is still unclear. Depending on the application stage, the biotransformation of Cu-based nanoparticles in soil may cause varied reactions. Therefore, the toxicity and reactivity of the metal or metal oxide Cu-based nanoparticles might differ from that of metallic copper [67, 76].

Gao and colleagues (2019) figured out that the maximum conversion of Cu-based nanoparticles to Cu ions was observed in acidic soils, and they investigated changes in the geochemical parameters of soils by taking twelve samples [30]. pH and organic matter of the soil affect the solubility and closure rate of copper oxide nanoparticles. Higher pH can effectively diminish the dissolution rate by inhibiting ion release. According to this study, the dissolving speed of copper oxide nanoparticles is different at soil pH 6.3 and below, for which there is a converse relation with the system dissolving speed in actual field situations. So, a high alkaline pH may cause the Cu nanoparticles not to dissolve easily compared to low pH (acidic soil), whereas due to this factor will release it more rapidly where it is applied. The experiment on the cupric oxide nanoparticle dissolution rate showed that the cupric oxide nanoparticle dissolution rate in soil with 7.2 pH was low compared to the 5.3 pH value [32].

Anderson and colleagues (2017) results also demonstrate how acidic pH (pH values less than 7) is a fundamental factor in regulating Cu bioavailability and mobility from soil into various ecosystems. This means that at pH 5.5, Cu dissolved 85% and 90%, as well as itself, while it did so with pH 7.0, pH 8.5, or both combined [9]. Consequently, Qiu and colleagues (2020) found that copper oxide nanoparticles had low dissolution in Lufa soil with a pH of 7.6. On the other hand, for the experiments carried out in calcareous soil with alkaline characteristics having a pH of 8.3, wheat growth was not affected by the integration of copper oxide nanoparticles at doses of 100 and 300 mg Cu/kg of soil. However, under an acidic environment characterized by a pH value of 4.8, crop toxicity occurred as a result of nanoparticle's toxicity impact on them. The change in soil pH will influence the behavior and destiny, including plant uptake and toxicity. The study found that the pH of the soil destabilized copper oxide nanoparticles. The zeta potential values of these particles shifted in response to increasing pH, as seen by their more negative charges. Nanoparticles' stability was reinforced, which increased their mobility and made it easier for them to reach deeper layers of the soil because the organic matter in the soil disintegrated and became a part of it. Lastly, copper oxide nanoparticles could pollute the soil. Also, when copper oxide nanoparticles reacted with H^+ ions present in acidic soils and released Cu^{2+} or $Cu(OH)^+$ that increased pH value, they caused higher electrical conductivity and decreased redox potential [65].

Another factor can be the duration of contact with Cu-based nanoparticles in the soil, as the concept is to assess their potential long-term effects on agricultural systems. In the soil, Cu-based nanomaterials could be changed or altered in other forms over time. Additionally, other environmental factors occurring during this period, such as weathering and erosion, may impact this. As pointed out by different researchers, soil variables produced more toxicities at three months old than copper sources. They also argued that some environmental parameters might cause more toxicity than nanotechnological properties or exposure doses. The weathering led Cu nanoparticles to be reduced to Cu_2O forms and subsidized to Cu_2S forms. The analyzed Cu species differed from each other in relation to weathered soils vs unweathered soils. It was also observed that with aging, proportions of Cu increased in treated soils like lettuce plants [62]. Because of the longer retention and exposure lengths, Cu nanoparticles would have plenty of chance to bind or interact with other compounds and pollutants prevalent in the soil, such as heavy metals, persistent pesticides, or their residues. The co-existence or co-contamination of Cu-based nanoparticles with different compounds and pollutants can have varying effects on the toxicity of the pollutant combination. This phenomenon is known as the Trojan horse-type effect [58]. Cu-based nanoparticles can easily chelate with contaminants and either reduce or counteract the toxicity in soils; due to the small size of these nanoparticles, they provide a broader contact surface and higher surface reactivity. Also, the parameters may shift the actual chemical structure of the contaminants and thus alter the mobility of the delivery medium, in this case, soil. Furthermore, Cu-based nanoparticles have the characteristic of tabling and carrying contaminants, which are absorbed by living things and bioconcentrated [58].

6 Future Research Directions

Based on the previous discussion, we state that integrating nanotechnology into agriculture will help to form a more sustainable future for the agricultural industry. Nanoscience and technology, as one of the forefront sciences of the twenty-first century, have the potential to bring substantial benefits to humanity that exceed the accompanying risks. They are guaranteed to promote sustainable agriculture development to meet the increasing global food demand. The prospect of cu-based nanoparticles in plant growth has attracted much attention due to the recent merging of modern agricultural science and nanotechnology systems. However, in the next years, more studies are expected to clarify the accurate physiological or molecular process that Cu-based nanoparticles stimulate and impact on plant cellular functions, like photosynthesis, nutrient uptake, and hormone regulation. Further researches are also required to enhance the useful aids on crops and diminish any negative impacts on plants and the environment by adjusting the concentration, size, and way of application of Cu-based nanoparticles. As the demand for food security increases, researchers are looking into the combined effects of mixing Cu-based nanoparticles with growth-promoting substances of crops. To fully harness the advantages of Cu-based nanoparticles for sustainable output, a comprehensive study method including genetics, nanotechnology, and plant physiology is essential. It is crucial to develop methods for producing Cu-based nanomaterials on a large scale and testing them in world agricultural settings. Presently, most research on these materials is either in the laboratory or controlled environments such as greenhouses within agriculture. However, the effectiveness of Cu-based nanomaterials needs to be evaluated under different climatic and environmental conditions by performing field trials on farms. They also involve complex and expensive procedures for their preparation. Therefore, before scaling up production and application, it is crucial to thoroughly refine the process and analyze costs to ensure its use in agriculture.

By considering the current status of development and constraints of nano-agriculture products, we suggest that future research emphasize the exact achievement of Cu-based nanoparticles against crop pests and pathogens, considering their influence on the fungus, insects, or bacteria reproduction, physiology, and behavior. Furthermore, due to increased concerns over antibiotic resistance in insects, combining Cu-based nanoparticles and other biocontrol agents may receive more attention in future studies. Extensive field trials must be conducted alongside long-term ecological monitoring to ascertain the practicability, safety, and economic feasibility of integrating Cu-based nanoparticles with traditional means of managing agricultural pests. Finally, it is essential to examine the potential contamination of these nanoparticles to adjacent surroundings, bearing in mind other organisms not targeted in the soil ecosystem due to possible effects caused by successive applications. As our dependence on nanotechnology develops worldwide, it is important to know the negative environmental effects of Cu nanoparticle materials. So, future research should be focused more on the consequences of Cu nanoparticles in soil ecosystems when they are engaged at the sub-microscopic level with plants, invertebrates, and soil bacteria. A multidisciplinary method incorporating ecology, toxicology, and nanotechnology will be crucial in guiding regulations and confirming the sustainable use of Cu-based nanomaterials without endangering the health and balance of our soil ecosystems. This gives the potential widespread use of Cu nanoparticles in various industries. Finally, just like with any recently established nanotechnology the safe application of Cu-based nanoparticles in agricultural areas will require clear laws and restrictions.

7 Conclusion

In conclusion, nanotechnology is presently being discovered for its potential to bring beneficial changes to the agricultural sector and other sciences. Sustainable agriculture is one of the most important trends for future agricultural development. Cu-based nanoparticles are significant nanomaterials because of their unique properties and wide technological applications. According to different recent studies, we conclude that Cu-based nanoparticles have promising and unique properties that can be used for various agricultural applications, including large surface area, excellent catalytic activity, tuneable properties, good electrical conductivity, and so on. The antibacterial, antifungal, and pesticide activities also make them a better candidate for use in combating plant diseases and are highly applicable in agriculture. Conventional or biological approaches can synthesize them. Compared with conventional synthetic methods, biological synthesis has the advantages of low toxicity and price, moderate reaction conditions, stable yields, and eco-friendliness. The physical and chemical synthetic methods remain expensive, involve sophisticated temperatures, and produce toxic by-products. Some recent reports showed that the safety of emerging nanomaterials is still uncertain. Literature suggests the long-term impact on the agricultural field and other biological entities should be investigated before using these nanomaterials extensively. Finally, several reports also showed the toxic effects of Cu-based nanoparticles on plants at high dosages/concentrations. According to some studies, higher concentrations may alter plant metabolism, which ultimately affects yield and is poisonous to other organisms in the ecosystem. Hence, for safer agricultural use of Cu-based nanoparticles for plant growth enhancements and nutrient delivery, a sensible comprehension of the interaction of nanoparticles with plants is achieved. To sum up, nanotechnology can be easily used to provide better yields with better quality. However, to fully use the promise of nanotechnology in agriculture and other environmental sciences, strict regulatory requirements must be implemented.

Competing Interests The authors declare that they have no competing interest.

Declaration of Generative AI and AI-Assisted Technologies in the Writing Process During the preparation of this work the book chapter author(s) have not taken any assistance from AI, and the book chapter author(s) reviewed and edited the content as needed and take(s) full responsibility for the content of the publication. All authors read and approved the final version of the manuscript.

References

1. Abbasi E et al (2016) Silver nanoparticles: synthesis methods, bio-applications and properties. Crit Rev Microbiol 42(2):173–180
2. Abd-Elsalam KA (2022) Copper-based nanomaterials: next-generation agrochemicals: a note from the editor. In: Copper nanostructures: next-generation of agrochemicals for sustainable agroecosystems. Elsevier, pp 1–14
3. Abd-Elsalam KA et al (2020) Copper-chitosan nanocomposite hydrogels against aflatoxigenic Aspergillus flavus from dairy cattle feed. J Fungi 6(3):112
4. Abd El-Aziz M et al (2019) Preparation and characterization of chitosan/polyacrylic acid/ copper nanocomposites and their impact on onion production. Int J Biol Macromol 123:856–865
5. Akintelu SA et al (2020) A review on synthesis, optimization, mechanism, characterization, and antibacterial application of silver nanoparticles synthesized from plants. J Chem 2020(1):3189043
6. Akintelu SA et al (2021) Green chemistry approach towards the synthesis of copper nanoparticles and its potential applications as therapeutic agents and environmental control. Current Res Green Sustain Chem 4:100176
7. Ameh T, Sayes CM (2019) The potential exposure and hazards of copper nanoparticles: a review. Environ Toxicol Pharmacol 71:103220
8. Anboo S et al (2022) Recent advancements in enzyme-incorporated nanomaterials: synthesis, mechanistic formation, and applications. Biotechnol Bioeng 119(10):2609–2638
9. Anderson A et al (2017) Soil chemistry influences the phytotoxicity of metal oxide nanoparticles. Int J Nanotechnol 14(1–6):15–21
10. Antony A, Boo J-H (2021) Size and shape selective metal oxide nanomaterials: preparation, characterization and prospective biomedical applications. In: Nanomaterials and Their Biomedical Applications, pp 83–105
11. Atwa AA et al (2017) Insecticidal effects of pure and silver-doped copper oxide nanosheets on Spodoptera littoralis (Lepidoptera: noctuidae). Can Entomol 149(5):677–690
12. Bachheti RK, Bachheti A, Husen A (Eds.) (2023) In: Nanomaterials for environmental and agricultural sectors. Springer, Singapore. https://doi.org/10.1007/978-981-99-2874-3
13. Bachheti RK, Bachheti A, Husen A (Eds.) (2024) In: Metal and metal-oxide based nanomaterials: singapore: synthesis, agricultural, biomedical and environmental interventions. Springer Nature, Singapore. https://doi.org/10.1007/978-981-99-7673-7

14. Bachheti RK, Bachheti A, Husen A (2025) Chitin-based nanoparticles for the agriculture sectors. Springer Nature Singapore Pte Ltd., 152 Beach Road, #21–01/04 Gateway East, Singapore 189721, Singapore
15. Bagherzadeh M et al (2022) Bioengineering of CuO porous (nano) particles: role of surface amination in biological, antibacterial, and photocatalytic activity. Sci Rep 12(1):15351
16. Bakshi M, Kumar A (2021) Copper-based nanoparticles in the soil-plant environment: assessing their applications, interactions, fate and toxicity. Chemosphere 281:130940
17. Balusamy SR et al (2023) Advancing sustainable agriculture: a critical review of smart and eco-friendly nanomaterial applications. J Nanobiotechnol 21(1):372
18. Bonthula S et al (2023) Recent advances in copper-based materials for sustainable environmental applications. Sustain Chem 4(3):246–271
19. Chen J et al (2019) Various antibacterial mechanisms of biosynthesized copper oxide nanoparticles against soilborne Ralstonia solanacearum. RSC Adv 9(7):3788–3799
20. Choudhary RC et al (2017) Cu-chitosan nanoparticle boost defense responses and plant growth in maize (Zea mays L.). Scient Reports 7(1):9754
21. Dimkpa CO et al (2015) Nano-CuO and interaction with nano-ZnO or soil bacterium provide evidence for the interference of nanoparticles in metal nutrition of plants. Ecotoxicology 24:119–129
22. Din MI, Rehan R (2017) Synthesis, characterization, and applications of copper nanoparticles. Anal Lett 50(1):50–62
23. Długosz O et al (2023) Cu_2O nanoparticles deposited on Y2O3 and CuO: synthesis and antimicrobial properties. J Cluster Sci 34(4):2153–2165
24. Dodevska T et al (2022) Application of biosynthesized metal nanoparticles in electrochemical sensors. J Serb Chem Soc 87(4):401–435
25. Dwivedi S et al (2016) Understanding the role of nanomaterials in agriculture. In: Microbial inoculants in sustainable agricultural productivity: Vol. 2: functional applications, pp 271–288
26. Ekanayake SA, Godakumbura PI (2021) Synthesis of a dual-functional nanofertilizer by embedding ZnO and CuO nanoparticles on an alginate-based hydrogel. ACS Omega 6(40):26262–26272
27. El-Abeid SE et al (2024) Antifungal activity of copper oxide nanoparticles derived from Zizyphus spina leaf extract against Fusarium root rot disease in tomato plants. J Nanobiotechnol 22(1):28
28. El-Saadony MT et al (2020) Ecofriendly synthesis and insecticidal application of copper nanoparticles against the storage pest Tribolium castaneum. Nanomaterials 10(3):587
29. Feregrino Pérez AA et al (2023) Interactions between nanomaterials and plant–microbe partnership. In: Agricultural and environmental nanotechnology: novel technologies and their ecological impact. Springer, pp 353–392
30. Gao X et al (2019) Effect of soil organic matter, soil pH, and moisture content on solubility and dissolution rate of CuO NPs in soil. Environ Sci Technol 53(9):4959–4967
31. Giannousi K et al (2013) Synthesis, characterization and evaluation of copper based nanoparticles as agrochemicals against Phytophthora infestans. RSC Adv 3(44):21743–21752
32. Guan X et al (2020) CuO nanoparticles alter the rhizospheric bacterial community and local nitrogen cycling for wheat grown in a calcareous soil. Environ Sci Technol 54(14):8699–8709
33. Gupta R et al (2024) Detection of vanillin in food products over Cu-laser induced graphene nanocomposite using the combined electrochemistry and UV–Vis spectroscopy principles. Microchem J 199:109984
34. Hermida-Montero L et al (2019) Aqueous-phase synthesis of nanoparticles of copper/copper oxides and their antifungal effect against Fusarium oxysporum. J Hazard Mater 380:120850
35. Hong J et al (2015) Toxic effects of copper-based nanoparticles or compounds to lettuce (Lactuca sativa) and alfalfa (Medicago sativa). Environ Sci Process Impacts 17(1):177–185
36. Hussain I et al (2016) Green synthesis of nanoparticles and its potential application. Biotech Lett 38:545–560
37. Azamal H, Bachheti RK, Bachheti A (Eds.) (2023) In: Current trends in green nano-emulsions: food, agriculture and biomedical sectors. Singapore, Springer Nature. https://doi.org/10.1007/978-981-99-5398-1

38. Husen A (2025) Plant response to silver nanoparticles. Springer Nature Singapore Pte Ltd., 152 Beach Road, #21–01/04 Gateway East, Singapore 189721, Singapore
39. Husen A (2024) Plant response to gold nanoparticles. Springer Nature Singapore Pte Ltd., 152 Beach Road, #21–01/04 Gateway East, Singapore 189721, Singapore
40. Husen A (2025) Emerging carbon nanomaterials for sustainable agricultural practices—synthesis, plant growth, performance, production and protection. Springer Nature Singapore Pte Ltd., 152 Beach Road, #21–01/04 Gateway East, Singapore 189721
41. John MS et al (2021) Biogenic synthesis of copper nanoparticles using bacterial strains isolated from an Antarctic consortium associated to a psychrophilic marine ciliate: characterization and potential application as antimicrobial agents. Mar Drugs 19(5):263
42. Kaphle A et al (2018) Nanomaterials for agriculture, food and environment: applications, toxicity and regulation. Environ Chem Lett 16:43–58
43. Kasana RC et al (2016) Copper nanoparticles in agriculture: biological synthesis and antimicrobial activity. Nanosci Food Agricul 3:129–143
44. Kora AJ (2022) Copper-based nanopesticides. In: Copper nanostructures: next-generation of agrochemicals for sustainable agroecosystems, pp 133–153
45. Kuang L et al (2015) Silicon nanoparticles synthesised through reactive high-energy ball milling: enhancement of optical properties from the removal of iron impurities. J Exp Nanosci 10(16):1214–1222
46. Kumar M et al (2024) A sustainable approach to prepare green synthesis of copper nanoparticles of Bauhinia variegata and Saussurea lappa: Unveiling in-vitro anti-obesity applications. Heliyon 10(8)
47. Lastovina TA et al (2016) Copper-based nanoparticles prepared from copper (II) acetate bipyridine complex. J Serb Chem Soc 81(7):751–762
48. León Jimenez E et al (2019) Síntesis y actividad insecticida de nanopartículas de Cu de Prosopis juliflora (Sw) DC y Pluchea sericea (Nutt.) sobre Phenacoccus solenopsis Tinsley (Hemiptera: Pseudococcidae). Revista de la Sociedad Entomológica Argentina 78(2):1–10
49. Leonardi M et al (2021) Smart nanocomposites of chitosan/alginate nanoparticles loaded with copper oxide as alternative nanofertilizers. Environ Sci Nano 8(1):174–187
50. Li X et al (2021) Characterization and apoptotic effect of copper nanoparticles biosynthesized from Ziziphus zizyphus leaf on human renal cell carcinoma A498 cells. Appl Nanosci 11:139–148
51. Li Y et al (2023) Application and mechanisms of metal-based nanoparticles in the control of bacterial and fungal crop diseases. Pest Manag Sci 79(1):21–36
52. Lobregas MOS, Camacho DH (2022) Green synthesis of copper-based nanoparticles using microbes. In: Copper nanostructures: next-generation of agrochemicals for sustainable agroecosystems, pp 17–44
53. Mahboub HH et al (2022) Protective effects of Allium hirtifolium extract against foodborne toxicity of Zinc oxide nanoparticles in common carp (Cyprinus carpio). Comp Biochem Physiol C: Toxicol Pharmacol 257:109345
54. Malviya T et al (2023) Gum Arabic capped Cu/Zn bimetallic nanoparticles for the germination and growth of chickpea. Mater Today: Proc
55. Mendez-Trujillo V et al (2022) Insecticidal effect of green bimetallic nanoparticles from Crotalaria longirostrata on Cotton Mealybug, Phenacoccus solenopsis. J Renw Mater 10(10):2543–2552
56. Mikula K et al (2020) Controlled release micronutrient fertilizers for precision agriculture–A review. Sci Total Environ 712:136365
57. Molahalli V et al (2024) Properties, synthesis, and characterization of Cu-based nanomaterials. In: Copper-based nanomaterials in organic transformations. ACS Publications, pp 1–33
58. Naasz S et al (2018) Environmental mixtures of nanomaterials and chemicals: the Trojan-horse phenomenon and its relevance for ecotoxicity. Sci Total Environ 635:1170–1181
59. Nasrollahzadeh M et al (2019) Recent developments in the plant-mediated green synthesis of Ag-based nanoparticles for environmental and catalytic applications. Chem Rec 19(12):2436–2479

60. Nzilu DM et al (2023) Green synthesis of copper oxide nanoparticles and its efficiency in degradation of rifampicin antibiotic. Sci Rep 13(1):14030
61. Peixoto S et al (2024) Responses of soil microbiome to copper-based materials (nano and bulk) for agricultural applications: an indoor-mesocosm experiment. NanoImpact 34:100506
62. Peng Cheng PC et al (2017) Fate and transformation of CuO nanoparticles in the soil-rice system during the life cycle of rice plants
63. Periakaruppan R et al (2023) Innovations in modern nanotechnology for the sustainable production of agriculture. ChemEngineering 7(4):61
64. Pothu R et al (2022) Vapour-phase selective hydrogenation of γ-Valerolactone to 2-methyltetrahydrofuran biofuel over silica-supported copper catalysts. Nanomaterials 12(19):3414
65. Qiu Y et al (2020) The aggregation and sedimentation of two different sized copper oxide nanoparticles in soil solutions: dependence on pH and dissolved organic matter. Sci Total Environ 731:139215
66. Rai M et al (2018) Copper and copper nanoparticles: Role in management of insect-pests and pathogenic microbes. Nanotechnol Rev 7(4):303–315
67. Rajput V et al (2020) Interaction of copper-based nanoparticles to soil, terrestrial, and aquatic systems: critical review of the state of the science and future perspectives. Rev Environ Contam Toxicol 252:51–96
68. Rashid MI et al (2023) Nanobiochar and copper oxide nanoparticles mixture synergistically increases soil nutrient availability and improves wheat production. Plants 12(6):1312
69. Saharan V et al (2015) Synthesis and in vitro antifungal efficacy of Cu–chitosan nanoparticles against pathogenic fungi of tomato. Int J Biol Macromol 75:346–353
70. Shafiq H et al (2024) Copper oxide nanoparticles induced growth and physio-biochemical changes in maize (Zea mays L.) in saline soil. Plants 13(8):1080
71. Shang H et al (2020) Copper sulfide nanoparticles suppress Gibberella fujikuroi infection in rice (Oryza sativa L.) by multiple mechanisms: contact-mortality, nutritional modulation and phytohormone regulation. Environ Sci: Nano 7(9):2632–2643
72. Shang H et al (2021) Copper oxide nanoparticle-embedded hydrogels enhance nutrient supply and growth of lettuce (Lactuca sativa) infected with Fusarium oxysporum f. sp. lactucae. Environ Sci Technol 55(20):13432–13442
73. Shende S et al (2015) Green synthesis of copper nanoparticles by Citrus medica Linn. (Idilimbu) juice and its antimicrobial activity. World J Microbiol Biotechnol 31:865–873
74. Shukla K et al (2024).Nanotechnology in sustainable agriculture: a double-edged sword. J Sci Food Agricul
75. Siddiqi KS, Husen A (2020) Current status of plant metabolite-based fabrication of copper/copper oxide nanoparticles and their applications: a review. Biomater Res 24(11):1–15. https://doi.org/10.1186/s40824-020-00188-1
76. Simonin M et al (2018) Negative effects of copper oxide nanoparticles on carbon and nitrogen cycle microbial activities in contrasting agricultural soils and in presence of plants. Front Microbiol 9:3102
77. Singh D et al (2023) Bacteria assisted green synthesis of copper oxide nanoparticles and their potential applications as antimicrobial agents and plant growth stimulants. Front Chem 11:1154128
78. Su C et al (2024) Copper-based nanomaterials: opportunities for sustainable agriculture. Sci Total Environ 171948
79. Sukumar S et al (2020) Green-synthesized rice-shaped copper oxide nanoparticles using Caesalpinia bonducella seed extract and their applications. ACS Omega 5(2):1040–1051
80. Sultana MJ et al (2023) Biosensing and anti-inflammatory effects of silver, copper and iron nanoparticles from the leaf extract of Catharanthus roseus. Beni-Suef Univer J Basic Appl Sci 12(1):26
81. Thakur S et al (2018) Green synthesis of copper nano-particles using Asparagus adscendens roxb. Root and leaf extract and their antimicrobial activities. Int J Curr Microbiol Appl Sci 7(4):683–694

82. Tortella G et al (2024) Copper nanoparticles as a potential emerging pollutant: divergent effects in the agriculture, risk-benefit balance and integrated strategies for its use. Emerging Contaminants 100352
83. Van Nguyen D et al (2022) Copper nanoparticle application enhances plant growth and grain yield in maize under drought stress conditions. J Plant Growth Regul 41(1):364–375
84. Vedeler EV (2017) In: Optical properties of metal nanoparticles investigated using polarised light. The University of Bergen
85. Velicogna JR et al (2020) Phytotoxicity of copper oxide nanoparticles in soil with and without biosolid amendment. NanoImpact 17:100196
86. Venkatesha N, Ramesh S (2018) Citric acid-assisted synthesis of nanoparticle copper catalyst supported on an oxide system for the reduction of furfural to furfuryl alcohol in the vapor phase. Ind Eng Chem Res 57(5):1506–1515
87. Vieillard J et al (2019) CuO nanosheets modified with amine and thiol grafting for high catalytic and antibacterial activities. Ind Eng Chem Res 58(24):10179–10189
88. Wassie AT, Ahmed IN, Bachheti A, Husen A, Bachheti RK (2024) Plant extract-assisted green synthesis and characterization of copper nanocomposite. In: Metal and metal-oxide based nanomaterials: synthesis, agricultural, biomedical and environmental interventions. Singapore, Springer Nature, pp 43–66
89. Wassie AT, Bachheti RK, Bachheti A (2023) Recent advances in nanoparticles for environmental monitoring and sensing: an overview. In: Nanomaterials for environmental and agricultural sectors. Singapore, Springer Nature, pp 107–122
90. Xiong T et al (2017) Copper oxide nanoparticle foliar uptake, phytotoxicity, and consequences for sustainable urban agriculture. Environ Sci Technol 51(9):5242–5251
91. Xiong T, Li S (2022) Effects, uptake, and translocation of Cu-based nanoparticles in plants. In: Toxicity of nanoparticles in plants, pp 131–170
92. Young M, Santra S (2014) Copper (Cu)–Silica nanocomposite containing valence-engineered Cu: a new strategy for improving the antimicrobial efficacy of Cu biocides. J Agric Food Chem 62(26):6043–6052
93. Yuvaraj M et al (2024) Slow release nanofertilizer for sustainable agriculture. In: Nanofertilizer delivery, effects and application methods. Elsevier, pp 53–69

Environmental Impacts of Copper-Based Nanomaterials in Agriculture

Marzieh Ghanbari Jahromi, Somayeh Anari Nobakht, and Zahra Hassanabadi

Abstract Copper based nanomaterials (Cu-NMs) are a promising tool for agriculture to increase plant productivity and resilience. Controlling the copper supply resulted in enhanced seed germination and vigorous seedling development and early stage growth. In addition, Cu-NMs also control abiotic stresses such as drought, salinity, and oxidative stress through enhanced antioxidant defense mechanisms and plant survival and yield. Potent antimicrobial properties were demonstrated, controlling plant diseases caused by bacteria, fungi, and viruses, and as a pesticide or fungicide, limiting the need for synthetic chemicals. In addition, Cu-NMs also enhanced soil microbial communities, creating an environment that is balanced and that contributes to long-term agricultural sustainability. However, environmental concerns include copper accumulation, potential toxicity, and soil microbial community disruption. The long-term effect of Cu-NMs on soil health, ecosystem balance, and food safety still needs further investigation. Sustainable practices and regulatory frameworks are therefore necessary for safe use in agriculture. The aim of this chapter is to provide an overview on environmental impacts of copper-based nanomaterials in agriculture.

Keywords Copper nanomaterials · Growth regulation · Soil microbial activity · Stress modulation

M. Ghanbari Jahromi (✉)
Department of Agricultural Science and Engineering, SR.C., Islamic Azad University, Tehran, Iran
e-mail: ghanbari@iau.ac.ir

S. A. Nobakht
Department of Horticultural Science, Karaj Branch, Islamic Azad University, Tehran, Iran

Z. Hassanabadi
Department of Medicinal Plants, Islamic Azad University, Shahrbabak Branch, Shahrbabak, Iran

R. K. Bachheti et al. (eds.), *Copper-Based Nanomaterial for Agricultural Practices*, Smart Nanomaterials Technology, https://doi.org/10.1007/978-981-96-8610-0_2

1 Introduction

Nanomaterials (NMs) have emerged as a new tool in agriculture, offering fresh approaches to soil management, insect management, and crop yield. In order to overcome agricultural issues, nanotechnology involves manipulating materials at the nanoscale (1–100 nm) to benefit from special qualities including improved absorption, controlled release, and greater reactivity. By enhancing nutrient delivery to the crop and offering higher efficiency than conventional fertilizers, nano fertilizers enhance crop development [10, 48].

Nano copper (Cu) and nano zinc fertilizers increase the bioavailability of essential micronutrients to give you healthier plants and higher yields. In addition, they decrease the environmental impact of excessive fertilizer use [49]. Highly effective pesticides and fungicides are NMs because of their antimicrobial properties. For example, copper nanoparticles (Cu-NPs) can stop the growth of bacteria, fungi, and pests that eat away at crops. Furthermore, NMs can be engineered as smart agrochemical delivery systems that introduce pesticides and herbicides in slow, controlled release [17].

The development of nano-structured soil conditioners can improve soil health and water retention and remediate contaminated soil. While their use is raising concerns about potential environmental and health risks. Accumulation of NPs in soils, water bodies, and plant tissues can cause unintended toxicity to non-target organisms such as beneficial soil microorganisms, aquatic life, and humans. The lack of comprehensive frameworks to deal with the unique properties of NMs presents regulatory and safety challenges in using NMs in agriculture. With the ongoing research in NMs, it will be important to find the right balance between what they can do for agriculture and strategies to minimize environmental and health impacts to ensure the safe and responsible use of nanotechnology in agriculture [10, 19].

2 Copper

Cu is a reddish-brown metal with atomic number 29 and reddish-brown color. It is well known for its high electrical and thermal conductivity and is thus used for electrical wiring and components. Cu's malleability, ductility, and corrosion resistance make it a common choice for use in many applications [34]. Because of its aesthetic appeal and durability, it is used in electrical wiring, plumbing, alloys, coinage, and construction. Cu is essential for life, humans and animals need small amounts of Cu to ensure proper enzyme and organ function. It also affects iron metabolism, nerve function and the production of red blood cells [18]. Ores in which Cu is found naturally include chalcopyrite and malachite and Cu can be found in its pure form, used by humans since prehistoric times [47].

2.1 Shape and Size of Copper-Based Nanomaterials

The potential for environmental cleanup has prompted the study of Cu-NMs. A variety of NMs have been made for use in a wide range of sizes, shapes, and porosities. The shape of these NMs is also strongly influenced by the circumstances and techniques of synthesis. Shape, size, and porosity all influence the overall removal yield of environmental pollutants by Cu-NMs, and hence, economic considerations and the commercialization of these innovative materials need to be considered [28].

The shape of Cu-NMs is considered an important structural property that has an eminent effect on its properties. The direct relationship between shape and size and the effects of shape on inherent properties like band gap energy can be observed [22]. Modulating reaction conditions can result in various shapes, including three-dimensional and three types of two-dimensional CuO NMs. Particle shape also produces differences in the specific surface area (SSA) of the particles. For example, the nano Cu is smaller, which can increase the degradation rate of methylene blue under visible irradiation. Nano-Cu is a potential visible light photocatalyst, particularly when smaller sized [28].

Size, surface area, morphology, and shape are all factors that significantly modify the properties of Cu-NBMs. These factors can significantly influence the optical and electrical properties of Cu-NMs. For instance, the widely available organic dye Coomassie brilliant blue R-250 displayed strong photocatalysis when exposed to rod-shaped CuO nanoparticles [36].

2.2 Crystalline Structure

Particles, fullerenes, nanoplates, and nanotubes are among the 36 primary classes into which NMs are divided. These materials are classified as either bulk materials or NMs based on their crystalline structure. The characteristics of NMs can be influenced by the proportion of material size to crystallite size, especially in applications like environmental cleanup. Catalytic efficiency may be adversely affected by crystalline structural defects [3, 28, 40]. Sreeju et al. [44] synthesized highly crystalline Cu nanoparticles with a high crystalline structure to increase their resistance to oxidation. This resistance is explained by the high crystalline materials' limited ability to absorb oxygen due to their lower surface energy. However, crystallinity hasn't had much of an impact on Cu-NMs' reactivity.

2.3 Zeta Potential

Electrokinetic or zeta potential is the potential at the slipping plane of a moving colloid particle under an electric field. Controlling electrostatic interactions and determining the stability of colloidal dispersions [26] is essential. The removal of environmental contaminants, such as adsorption or advanced oxidation processes, depends upon a medium's initial pH. The removal efficiency increases with increasing pH from 2.0 to 6.0 when the medium is acidic. Nevertheless, the removal efficiency decreases from 6.5 to 10 pH due to the lower adsorption tendency of particles with negative charges toward 2 chlorophenol [28]. The adsorption of special contaminants also is affected by the surface charge of NPs. The surface charge of CuO NPs was used by [31] to remove phosphate, with a 23.9 mg phosphate/g CuO removal efficiency for the phosphate.

2.4 Optical Properties

Cu-NMs are catalytic materials utilized in many different applications, and their performance depends on their optical characteristics [42]. When compared to traditional NMs, Cu-NMs can exhibit increased efficiency because of their narrow band gap. Recent studies have shown that visible light absorption and photo-generated electron migration can be promoted through loading Cu_2O nanospheres and nano-chitosan into a 3DrGO network, leading to the effective degradation of rhodamine B (RhB) dye [28].

Cu-NMs have also been shown to enhance the optical performance of multi-component catalysts. To remove congo red dye, graphene, which is becoming increasingly popular for cleaning polluted water and wastewater, was utilized as a supporting material for a multi-component structure using polyaniline cuprous oxide. Cu2O and graphene components have predicted band gaps of 1.62 and 2.04 eV, respectively. Various UV–Vis irradiations may be made adsorbent by mixing Cu-NMs with other metal and metal oxide NMs [28].

Combining Cu-NMs with additional metal oxide NMs can improve the photocatalytic effectiveness of the obtained characteristics. For instance, when exposed to visible sunlight, acid red B aqueous solution was decolored using a p-n heterojunction between Cu_2O and TiO_2 NMs. When compared to pure-phase TiO2 NMs, this combination significantly increased the Cu_2O/TiO_2 visible light photocatalytic efficiency [29].

Additionally, CuO has been effectively combined with other metal oxides to improve the removal of complex organic molecules. In order to remove benzophenone (BP) from the contained solution, Esmaili Hafshejani and Nezamzadeh-Ejhieh [14] investigated the photocatalytic activity of the coupled ZnO-CuO and ZnS-CuS semiconductors supported onto clinoptilolite nanoparticles (CNP) and micronized one (CMP). Simamora et al. [43] performed a simultaneous photocatalytic removal of dyes (methylene blue) and production of solar fuel (H_2) by splitting the water and seawater in the presence of CuO supported nano-sized TiO_2 under UV–visible irradiation. They achieved up to 100% nitrate removal and hydrogen generation up to 14 μmol/L of H_2. Cu-NMs have also been used to remove higher initial concentrations of nitrate from contaminated solutions.

3 Enhancing Seed Germination by Copper Nanomaterials

Nano stimulants, a class of nanoparticles aimed at boosting seed performance, have demonstrated impressive promise in boosting seed germination, early seedling growth, and overall crop productivity. Metal-based nanoparticles, carbon-based NMs, and nano-formulated fertilizers and hormones affect a number of biochemical and physiological processes in seeds to increase germination rates and vigor under normal and stress conditions [45].

Mechanisms of action are increased water uptake and seed hydration, stimulation of metabolic activity, increased nutrient availability, and protection against oxidative stress. Enzymes such as amylase, protease, and lipase that break down stored food reserves in the seed (stored food reserves) to provide energy for embryo growth and development [23] can be made more active by NMs. Nano formulated nutrients, like nano fertilizers, make essential minerals more available at germination time and help with better early root and shoot development. Plant hormones like gibberellins and auxins are important to seed germination, and NMs can also affect this [30, 45].

Another benefit of nano-stimulants, and especially those that are antioxidant, is protection from oxidative stress. Carbon nanotubes (CNTs) have been shown to promote seed germination by increasing water uptake and enhancing nutrient absorption [2], as have silica nanoparticles (SiO2-NPs) in improving plant resilience to stress and root growth [2]. ZnO-NPs increase germination by stimulating enzyme activity in seeds [7], including amylase and auxin [7].

Cu-NPs are used to enhance crop seed germination by enhancing water uptake and accelerating metabolic and enzyme activities. They also trigger antioxidant defense systems, decreasing oxidative stress under abiotic stress conditions. The result is faster, more uniform germination, healthier, more vigorous seedlings, and better overall plant health [45]. In vitro, biogenic ZnO and Cu NPs were used by [37] on black gram seeds. These findings indicated that seeds treated with different concentrations of NPs (600 mg/l ZnO and 400 mg/l Cu) significantly improved seed quality and thus have the potential to enhance germination and seedling vigor (Table 1).

4 Increasing Plant Growth and Productivity by Copper-Based Nanomaterials

Cu-based NMs are essential micronutrients for plant growth photosynthesis and respiration. 30 to 50 percent of crop yield is due to commercial fertilizers and nano-sized nutrients that can enhance nutrient release, increase efficiency, lower waste and decrease environmental impact. Nano fertilizers deliver nutrients directly to plants in order to encourage plant growth and increase crop yields. Nano fertilizers are superior to conventional fertilizers in terms of fertilizer utilization, reduction of nutrient losses and environmental impact. In several studies, copper-NMs have been proven to improve crop nutritional status, reduce disease incidence, and increase

Table 1 The effect of copper nanomaterials on germination and plant growth

Author	Plant	Results
[35]	Chick pea	Gum acacia (GA) capped Cu-Ag bimetallic nanoparticles (Cu/Ag@GA2) were synthesized using hydrazine hydrate assisted co-reduction of copper sulfate and silver nitrate. The nanoparticles had a porous network structure with small needle-shaped bud-like structures. They showed good shelf life and could be stored for over 140 days. Cu/Ag@GA2 positively affected the germination and growth of *Cicer arietinum* L. seeds, with an optimum concentration of 200 mg/L
[27]	Barely	Low concentrations of Cu-NPs and CuO NPs enhanced seed germination and growth parameters, while concentrations above 500 mg/L displayed inhibiting effects. CuO NPs treatment resembled $CuSO_4$, mainly acting via released Cu ions. However, Cu-NPs exhibited different behavior, as they contributed together with Cu ions in barley response to Cu-NPs
[15]	Wheat	Biogenic Cu-NPs were synthesized from mangrove leaves, *Avicennia marina* (CE). The nanoparticles were primarily composed of D-mannose, D-fructose, L-sorbose, D-galactose, and D-talose. The nanoparticles were found to have a carbohydrate and phenolic content of 35.6% and 3.13mgGA/g, respectively. The impact of the biogenic Cu-NPs on wheat seedling growth was dose-dependent. Treatment with 0.06 mg/mL of Cu-NPs promoted growth by 172.78 and 215.94 ± % for wheat root and shoot, respectively. However, the lowest relative growth percentage was recorded when applying 0.43 mg/mL of Cu-NPs. The uptake of copper was higher in the root than in the shoot, suggesting the accumulation of the nanoparticles
[37]	Blackgram	The use of biogenic ZnO and Cu-NPs in in vitro invigoration studies on black gram seeds. The results show that seeds treated with different concentrations of nanoparticles, including 600 mg/l ZnO and 400 mg/l Cu, significantly improved seed quality, demonstrating their potential for improving germination and seedling vigor
[8]	Tomato	The study investigated the impact of cuprous oxide nanoparticles (Cu_2O NPs) from *Flacourtia montana* leaf extracts on tomato seed germination, growth, and vigor index. The results showed a dosage-dependent response, with the highest germination percentage (95%) observed at 20ppm Cu_2O NPs. Tomato seedlings showed increased root and shoot elongation up to 20ppm Cu_2O NPs concentration, but further concentrations negatively affected growth and development. Leaf pigments showed an increased trend in tomato plants after Cu_2O NPs treatment, with lower chlorophyll a, b, and carotenoid pigments. The study also found increased activity of antioxidant enzymes, such as Superoxide Dismutase (SOD) and Glutathione Peroxidase (GPX), with increased lipid peroxidation and decreased free radical scavenging activity
[24]	*Citrus reticulata*	The study explores the green synthesis of nanoparticles using plants, specifically zinc and copper nanoparticles. Green nanoparticles enhanced in vitro germination parameters due to their low toxicity and high efficacy. Zinc nanoparticles showed significant results in germination parameters, while copper nanoparticles showed higher toxicity and triggered the production of antioxidative enzymes and non-enzymatic compounds

crop yield and food safety. The effects of Cu-NMs depend on concentration and dose. Cu-NMs can modify plant photosynthesis, respiration, and mineral uptake, altering chlorophyll content, chloroplast structure, enzymes, mitochondrial membranes, and energy production [27].

4.1 Photosynthesis Rate

It has been found that cu-nms enhance photosynthesis rates in plants by optimizing several physiological, biochemical, and molecular processes. Cu is a key micronutrient participating in the photosynthetic electron transport chain, as electron transfer occurs between photosystem I (PSI) and photosystem II (PSII). Cu-NMs can improve the efficiency of these electron transport channels for more effective light energy collection and conversion during photosynthesis by increasing Cu availability at the cell level [11, 33].

In addition to having a significant impact on light absorption and photosynthesis, CU-NPs also play a key role in chloroplast integration. It can enhance the photosynthetic process and assist plants in maintaining chlorophyll content. Cu-NPs also significantly increased the antioxidant systems in plants, such as superoxide dismutase-inducing antioxidant enzymes. This resulted in the enhancement of plants' potential in several ways. Cu-NMs indirectly impact both stomatal conductance and the plant's gas exchange with its surroundings. Better regulating stomatal opening and shutting can boost photosynthetic rates and improve CO_2 absorption into leaves [6, 15].

4.2 Respiration

Cu-NMs can stimulate tuber plant respiration rates by increasing metabolic activities linked to cell energy generation. Cu is needed in those enzymes that are related to the transport of electrons in mitochondria, the last one within the sequence being cytochrome c oxidase. Cu-NMs enhance the bioavailability of Cu ions, thus enhancing electron transport in mitochondria and increasing power output by oxidative phosphorylation [12].

The last complex of eukaryotic oxidative phosphorylation in mitochondria, cytochrome c oxidase, is one enzyme associated with Cu-NPs that might be improved. Cu-NMs can also improve a plant's resistance to various environmental factors, including drought and salt. In these strange conditions, it helps plants survive [50].

4.3 Minerals Uptake

Cu-NMs enhance the movement of ions by allowing their passage across cell membranes. This has several roles, including light capture for photosynthesis, the activation of enzymes, regulation of the element's osmotic balance, and nutrient equilibrium. They also enhance the general absorption of the macronutrients, more so the iron, zinc, and micronutrient density. The Cu-NMs allow plants to balance nutrient uptake following gene expression, keeping nutrient intakes and availability beyond toxic and shortage levels [51].

Cu-NMs aid plants in coping with abiotic stress conditions by enhancing nutrient use efficiency. They increase the absorption and translocation of nutrients from roots to shoots, providing adequate mineral supply to the growth and metabolic functions under stress. Overall, Cu-NMs increase healthier, more resilient plants, with enhanced access to essential mineral nutrients, and thus enhance growth, yield, and stress tolerance [46].

5 Modulation of Abiotic Stress in Plants by Copper-Based Nanomaterial

An important drawback of plant growth and crop yield enhancement is abiotic stress, which includes drought, salinization, high temperatures, and heavy metal toxicity. Thus, the efficacy of Cu-NMs to enhance plant resistance to abiotic stresses has been investigated [32].

These NMs can regulate ROS, promote nutrient uptake, and activate stress-responsive pathways. Stimulation of the plant's antioxidant defense system (which reduces oxidative stress and cellular damage) provides the means to regulate ROS. The increased uptake of essential nutrients such as nitrogen and phosphorus also improves nutrient uptake and root development. Under stress, stress-responsive genes are activated to stabilize proteins, maintain cellular osmotic balance, and improve water retention [39].

Cu-NMs can also protect photosynthetic machinery by maintaining chlorophyll stability and improving electron transport in photosystems I and II. Drought stress leads to water deficit, which affects various physiological and metabolic processes in plants. Cu-NMs promote root elongation, facilitating deeper water absorption, and enhancing the synthesis of osmoprotectants like proline [5]. High salinity stress causes ionic and osmotic stress, preventing growth and productivity. Cu-NMs can induce the synthesis of heat shock proteins (HSPs) to stabilize proteins and prevent heat-induced damage. The application of Cu-NMs in agriculture has the potential to revolutionize plant stress management. However, the concentration and size of NPs are critical factors in determining their efficacy and potential toxicity [16]. Table 2 shows some related studies on Cu-NMs under different abiotic stress.

Table 2 The effect of copper nanomaterials on modulating abiotic stress in plants

Author	Plant	Results
[13]	Wheat	This study investigated the physiological and biochemical effects of polyvinylpyrrolidone (PVP)-coated Cu nanoparticles (PVP–Cu NP) in Triticum aestivum under salt and drought stress. The results showed that PVP-Cu NP treatments, especially the N1 dose, were effective in restoring growth rate and regulating water relations. They also reduced oxidative damage by stimulating antioxidant enzymes superoxide dismutase, peroxidase, and ascorbate peroxidase. The N1 dose PVP-Cu NP application provided stress tolerance and limited copper-induced adverse effects under all stress conditions. The findings suggest that PVP-Cu NP applications can be a valuable tool for minimizing adverse effects and improving the effectiveness of copper nanoparticles in agricultural applications
[39]	Wheat	A study on wheat crop growth during critical growth stages (CGS) found that drought significantly decreased yield. However, Cu-NPs application improved wheat growth and yield in various aspects. Principal component analysis revealed that DS during the anthesis stage had the most significant negative impact on yield. Applying Cu-nanoparticles at 300 mg L^{-1} was an effective strategy for improving crop productivity
[32]	Common bean	This study investigates the effects of potassium copper sulfate hydrate nanoparticles (PCSH NPs) on the growth and biochemical characteristics of common bean plants under salinity stress. The results show that PCSH NPs can modulate salt stress by increasing growth, chlorophyll content, relative water content, and proline content while decreasing MDA, EL, and enzyme activities. The study suggests that seed priming and soil application of PCSH NPs could help adjust the adverse effects of salinity on common bean plants
[16]	Wheat	The study uses Cu-chitosan nanoparticles (Cu-CNP) as eco-friendly priming agents to induce salt and PEG-induced hyperosmotic stress tolerance in wheat seedlings. Seed priming is a simple technique that minimizes soil fertility and requires a minimal priming agent. The study found that Cu-CNP priming enhanced POD activity under salt stress but decreased it under PEG-induced hyperosmotic stress. However, it also increased β-carotenoids, total carotenoids, chlorophyll a, b, and total chlorophyll under normal and stress conditions. The stress-insulating potential of Cu-CNP was reinforced by enhanced plant and photosynthetic pigment production

6 Soil Environment

Crop growth depends on soil, which is an interface for micro and macro flora, nutrients, organic matter, humic acids, and water. Cu-NMs can indirectly promote soil health by enhancing nutrient utilization, but they are also susceptible to toxic enrichment, which may adversely impact non-target microorganisms in the soil. Consequently, these NMs require careful consideration [9].

6.1 Copper-Based Nanomaterials on Soil Physicochemical Properties

Cu-NMs can change soil physicochemical properties to a significant extent and alter plant growth, nutrient availability and soil health. These effects depend on the interaction between Cu-NMs and soil components such as organic matter, minerals, and microorganisms [45].

Cu-NMs influence soil pH and acidic soils to affect nutrient solubility. Proper management is necessary to avoid undesirable pH shifts that may affect nutrient availability. Moreover, Cu-NMs can change the availability of essential nutrients, including copper, a crucial micronutrient for plant uptake. Despite that, high copper concentrations are toxic to plants and soil microorganisms and can impair nutrient uptake [20].

Cu-NMs can indirectly influence soil structure and porosity through Cu-NMs mediated aggregation of soil particles to improve water retention and drainage. Nevertheless, overuse could break soil aggregates, resulting in compacted or poorly structured soils. Cu-NMs may affect cation exchange capacity (CEC) and further contribute to the pool of available cations. Cu-NMs may have a moderate presence in the soil and improve its CEC but high concentrations may compete with other essential nutrients and displace them. Cu-NMs can affect microbial activity and organic matter decomposition; low concentrations stimulate microbial activity while high concentrations inhibit processes. Cu-NM also affects water retention and drainage, with improved water retention promoting plant growth during dry periods [9, 45].

6.2 Copper-Based Nanomaterials on Soil Microbial Communities

Both the positive and negative effects of Cu-NMs on soil microbial communities are critical for maintaining ecosystem balance. Bacteria, fungi, and archaea are part of these communities, which are important for nutrient cycling, decomposition of organic matter, and overall health of the soil. Excess concentrations of Cu-NMs, however, can be toxic, perturbing microbial processes and community structure [41].

Cu-NMs can modify microbial diversity since they are necessary for many enzymatic processes. $Cu2^{+}$ concentrations at high levels can be toxic, killing off sensitive species and allowing more resistant strains to thrive. They can also damage soil fertility. According to [45], bacteria are especially vulnerable to variations in copper levels. Cu-NMs can harm bacterial cell membranes, interfere with enzyme activity, and generate ROS (overproduction), which results in oxidative stress.

Fungi communities, such as mycorrhizal fungi, play a crucial role in forming symbiotic relationships with plants, aiding in nutrient uptake. Cu-NMs can affect

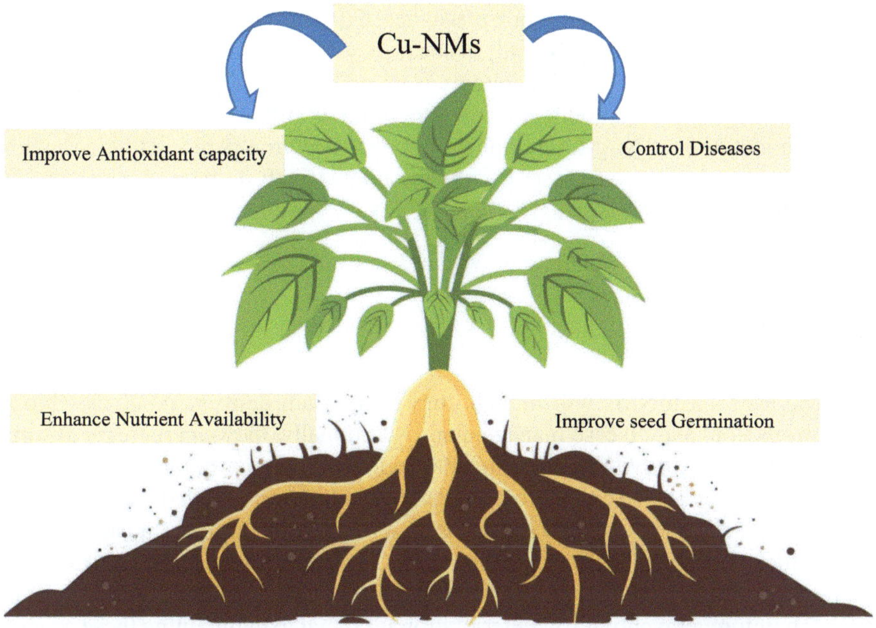

Fig. 1 The effects of copper-based nanomaterial on different aspects of agriculture

fungal communities by enhancing nutrient acquisition but can inhibit growth and disrupt symbiotic relationships, leading to reduced nutrient exchange [38].

Enzymatic activities, such as phosphatases, dehydrogenases, and ureases, can also be disrupted by Cu-NMs. Nitrogen cycling, particularly nitrification and denitrification, can be influenced by Cu-NMs, with nitrifying bacteria sensitive to high concentrations of copper. Oxidative stress is another significant negative effect of Cu-NMs on soil microbes, as the release of $Cu2^{+}$ ions can generate excessive reactive oxygen species, damaging microbial DNA, proteins, and lipids [4, 38]. Figure 1 shows the impacts of Cu-NMs in agriculture.

7 Concerns of Copper-Based Nanomaterials in Agriculture

Cu-NMs have the potential both to benefit and concern agriculture. However, they can increase crop productivity and protect plants from diseases, but they can also be phytotoxic, disrupt soil microorganisms, bioaccumulate, contaminate the environment, and develop resistance. Plant growth is hindered, photosynthesis is reduced, and cellular damage is caused by the excessive accumulation of Cu-NMs [1]. It can also upset soil nutrient cycles and decrease soil fertility over time. This can contaminate water sources and release Cu-NMs into the environment, thereby posing risks to aquatic ecosystems. Overuse of Cu-NMs as antimicrobial agents can create new diseases, as is the case with antibiotic resistance. Cu-NMs have not been investigated for their long-term reactivity in soil, instead, they focus on the diseases they may cause to the soil's health and inhabitants from an ecological perspective. Furthermore, the mitigation of nanomaterials in agriculture is still in its infancy, and it may yet have undetermined long-term implications on the environment and human health. Given these worries, it is necessary to conduct suitable analysis, control, and research on the safe use of Cu-NMs in agriculture [9, 45].

8 Conclusions

Applications for Cu-NMs will include improving plant delivery methods, abiotic stress protection, and disease management. Additionally, they can increase absorption, transport certain vital microminerals, and strengthen an organ's resistance to stress. However, it is also crucial to consider their relationship with the environment. Phytotoxicity occurs when pests are misused or improperly applied, causing antagonism in the soil microbiota or polluting water. Copper pollution of the environment, food safety, and bioaccumulation are some of the drawbacks of using copper nanoparticles. Therefore, preventing soil deterioration is crucial to maintaining these ecosystems' long-term equilibrium. Cu-NM-related applications witnessed an uptrend in agricultural applications such as seed coating Technology, incorporated fertilizers, pesticide-fungicide management, abiotic stress tolerance, improvement of SM content in positive ways, disease control, SRF, and biodegradable coatings. Cu-NMs are a rich source of biologically active copper ions required for enzyme activation, photosynthesis, and other metabolic processes necessary for seed germination and the first stages of plant development. Since they are also antimicrobial, they also serve as an antimicrobial control and pesticide/fungicide alternative. They enhance plant capacity to withstand biotic factors and diminish chemical fertilizers' use. Also, they conserve the desirable microorganisms that enhance soil nutrient cycling and profile formation. Cu-NMs can also be used in low-release fertilizers—which supply a steady and constant dosage of copper. This also avoids copper leaching and has long-term advantages for crops because they are eco-friendly coatings for fertilizers and pesticides. Moreover, the prospect of utilizing Cu-NMs in agriculture might also promote plant pathogens' ability to reduce these particles' antimicrobial capability. However, Cu-NMs can be used to improve growth-promoting conditions and decrease abiotic stress and disease burden in plants in agricultural applications, together with an understanding of environmental impacts. Furthermore, even though Cu-NMs are non-toxic, their long-term impacts must be thoroughly established before the best way to include them in agricultural systems and practices with long-term consequences can be studied, regulated, and carried out.

Competing Interests The authors declare that they have no competing interest.

Declaration of Generative AI and AI-Assisted Technologies in the Writing Process During the preparation of this work the book chapter author(s) have not taken any assistance from AI, and the book chapter author(s) reviewed and edited the content as needed and take(s) full responsibility for the content of the publication. All authors read and approved the final version of the manuscript.

References

1. Abd-Elsalam KA (2022) Copper-based nanomaterials: Next-generation agrochemicals: a note from the editor. In: Copper nanostructures: next-generation of agrochemicals for sustainable agroecosystems. Elsevier, pp 1–14
2. Abdoli S, Ghassemi-Golezani K (2023) Emerging roles of nanomaterials in plant-salinity interaction. In: New frontiers in plant-environment interactions: innovative technologies and developments, pp 93–119
3. Adamu A, Isaacs M, Boodhoo K, Abegao FR (2023) Investigation of Cu/TiO_2 synthesis methods and conditions for CO2 photocatalytic reduction via conversion of bicarbonate/carbonate to formate. J CO_2 Utilization 70:102428
4. Adhikari K, Timilsina A, Chen H (2024) The effects of size and surface-coating of CuO-nanoparticles on extractable Cu and enzyme activities in soil. Soil Environ Health 2(1):100065
5. Ahmed F, Javed B, Razzaq A, Mashwani ZUR (2021) Applications of copper and silver nanoparticles on wheat plants to induce drought tolerance and increase yield. IET Nanobiotechnol 15(1):68–78
6. Alhaithloul HAS, Ali B, Alghanem SMS, Zulfiqar F, Al-Robai SA, Ercisli S, Abeed AH (2023) Effect of green-synthesized copper oxide nanoparticles on growth, physiology, nutrient uptake, and cadmium accumulation in Triticum aestivum (L.). Ecotoxicol Environ Safety 268:115701
7. Amer HE (2024) Green synthesized ZnO Nanoparticles Improve the growth and phytohormones biosyn-thesis and modulate the expression of resistance genes in phaseolus vulgaris. Egypt J Bot 64(1):145–163
8. Ananda S, Shobha G, Shashidhara KS, Mahadimane V (2019) Nano-cuprous oxide enhances seed germination and seedling growth in Lycopersicum esculentum plants. J Drug Delivery Therapeutics 9(2):296–302
9. Bakshi M, Kumar A (2021) Copper-based nanoparticles in the soil-plant environment: assessing their applications, interactions, fate and toxicity. Chemosphere 281:130940
10. Balusamy SR, Joshi AS, Perumalsamy H, Mijakovic I, Singh P (2023) Advancing sustainable agriculture: a critical review of smart and eco-friendly nanomaterial applications. J Nanobiotechnol 21(1):372
11. Barreto DM, Tonietto AE, da Silva Oliveira EWR, Amaral CDB, Pulgrossi RC, Polpo A, Lombardi AT (2023) Photosynthesis-related responses of Chlorella sorokiniana (Trebouxiophyceae) to environmentally relevant concentrations of Cu nanoparticles. Phycologia 62(4):315–323
12. Da Costa MVJ, Kevat N, Sharma PK (2020) Copper oxide nanoparticle and copper (II) ion exposure in Oryza sativa reveals two different mechanisms of toxicity. Water Air Soil Pollut 231:1–16
13. Ekim R, Arikan B, Alp-Turgut FN, Koyukan B, Ozfidan-Konakci C, Yildiztugay E (2024) Polyvinylpyrrolidone-coated copper nanoparticles dose-dependently conferred tolerance to wheat under salinity and/or drought stress by improving photochemical activity and antioxidant system. Environ Res 241:117681
14. Esmaili-Hafshejani J, Nezamzadeh-Ejhieh A (2016) Increased photocatalytic activity of Zn (II)/Cu (II) oxides and sulfides by coupling and supporting them onto clinoptilolite nanoparticles in the degradation of benzophenone aqueous solution. J Hazard Mater 316:194–203
15. Essa HL, Abdelfattah MS, Marzouk AS, Shedeed Z, Guirguis HA, El-Sayed MM (2021) Biogenic copper nanoparticles from Avicennia marina leaves: Impact on seed germination,

detoxification enzymes, chlorophyll content and uptake by wheat seedlings. PLoS ONE 16(4):e0249764
16. Farooq T, Nisa ZU, Hameed A, Ahmed T, Hameed A (2022) Priming with copper-chitosan nanoparticles elicit tolerance against PEG-induced hyperosmotic stress and salinity in wheat. BMC chemistry 16(1):23
17. Francis DV, Abdalla AK, Mahakham W, Sarmah AK, Ahmed ZF (2024) Interaction of plants and metal nanoparticles: Exploring its molecular mechanisms for sustainable agriculture and crop improvement. Environ Int 108859
18. Frydrych A, Krośniak M, Jurowski K (2023) The role of chosen essential elements (Zn, Cu, Se, Fe, Mn) in food for special medical purposes (FSMPs) dedicated to oncology patients—critical review: state-of-the-art. Nutrients 15(4):1012
19. Gao J, Zhu Y, Zeng L, Liu X, Yang Y, Zhou Y (2024) Recent advances on environmental behavior of Cu-based nanomaterials in soil-plant system: a review. J Environ Manage 361:121289
20. Gao M, Chang J, Wang Z, Zhang H, Wang T (2023) Advances in transport and toxicity of nanoparticles in plants. J Nanobiotechnol 21(1):75
21. Gomes DG, Pieretti JC, Lourenço IM, Oliveira HC, Seabra AB (2022) Copper-based nanoparticles for pesticide effects. Inorganic Nanopesticides and nanofertilizers: a view from the mechanisms of action to field applications. Springer International Publishing, Cham, pp 187–212
22. Gomes SI, Roca CP, Pegoraro N, Trindade T, Scott-Fordsmand JJ, Amorim MJ (2018) High-throughput tool to discriminate effects of NMs (Cu-NPs, Cu-nanowires, $CuNO_3$, and Cu salt aged): transcriptomics in Enchytraeus crypticus. Nanotoxicology 12(4):325–340
23. Gomes-Junior FG, Rohr LA, Henning FA (2022) Nanoparticles on seed performance. Inorganic nanopesticides and nanofertilizers: a view from the mechanisms of action to field applications. Springer International Publishing, Cham, pp 103–122
24. Hussain M, Raja NI, Mashwani ZUR, Iqbal M, Ejaz M, Yasmeen F, Sohail (2017) In vitro germination and biochemical profiling of Citrus reticulata in response to green synthesised zinc and copper nanoparticles. IET Nanobiotechnol 11(7):790–796
25. Jahan I, Erci F, Isildak I (2021) Facile microwave-mediated green synthesis of non-toxic copper nanoparticles using Citrus sinensis aqueous fruit extract and their antibacterial potentials. J Drug Delivery Sci Technol 61:102172
26. Jahan MN, Alam MA, Rahman MM, Hoque SM, Ahmad H (2024) Mesoporous Fe_3O_4/SiO_2/ poly (2-carboxyethyl acrylate) composite polymer particles for pH-responsive loading and targeted release of bioactive molecules. RSC Adv 14(32):23560–23573.
27. Kadri O, Karmous I, Kharbech O, Arfaoui H, Chaoui A (2022) Cu and CuO nanoparticles affected the germination and the growth of barley (Hordeum vulgare L.) seedling. Bullet Environ Contaminat Toxicol 108(3):585–593
28. Khalaj M, Kamali M, Khodaparast Z, Jahanshahi A (2018) Copper-based nanomaterials for environmental decontamination–an overview on technical and toxicological aspects. Ecotoxicol Environ Saf 148:813–824
29. Kute AD, Gaikwad RP, Warkad IR, Gawande MB (2022) A review on the synthesis and applications of sustainable copper-based nanomaterials. Green Chem 24(9):3502–3573
30. Landa P (2021) Positive effects of metallic nanoparticles on plants: overview of involved mechanisms. Plant Physiol Biochem 161:12–24
31. Mahdavi S, Akhzari D (2016) The removal of phosphate from aqueous solutions using two nano-structures: copper oxide and carbon tubes. Clean Technol Environ Policy 18:817–827
32. Naghavi F, Khoshroo SMR, Kazemipour M, Zarandi MM (2023) Potassium copper sulfate hydrate nanoparticles modulated salinity stress through improving germination, growth, and biochemical attributes of common bean (Phaseolus vulgaris L.). Russian J Plant Physiol 70(6):152
33. Pandey N, Patel A, Tiwari S, Prasad SM (2022) Differential response of copper nanoparticles and ionic copper on growth, chlorophyll fluorescence, oxidative stress, and antioxidant machinery of two paddy field cyanobacteria. Ecotoxicology 31(6):933–947

34. Polyakov B, Novikovs A, Leimane M, Kadiwala K, Zubkins M, Butanovs E, Vlassov S (2023) Comparison of the resistivities of nanostructured films made from silver, copper-silver and copper nanoparticle and nanowire suspensions. Thin Solid Films 784:140087
35. Prabha M, Malviya T, Kumar A, Singh V (2023) Effect of Gum acacia capped Cu-Ag bimetallic nanoparticles on germination and growth of gram seeds (Cicer arietinum L.). Mater Today: Proc
36. Pérez-Hernández H, García-Mayagoitia S, Torres-Gómez PA, Campos-Montiel RG, Fernández-Luqueño F (2022) Ecological effects of copper NPs: advantages and drawbacks regarding current and potential applications. In: Copper nanostructures: next-generation of agrochemicals for sustainable agroecosystems. Elsevier, pp 719–750
37. Raja K, Sowmya R, Sudhagar R, Moorthy PS, Govindaraju K, Subramanian KS (2019) Biogenic ZnO and Cu nanoparticles to improve seed germination quality in blackgram (Vigna mungo). Mater Lett 235:164–167
38. Rajput V, Minkina T, Ahmed B, Sushkova S, Singh R, Soldatov M, Gorovtsov A (2020) Interaction of copper-based nanoparticles to soil, terrestrial, and aquatic systems: critical review of the state of the science and future perspectives. Rev Environ Contam Toxicol 252:51–96
39. Raza MAS, Amin J, Valipour M, Iqbal R, Aslam MU, Zulfiqar B, Elsalahy HH (2024) Cu-nanoparticles enhance the sustainable growth and yield of drought-subjected wheat through physiological progress. Sci Rep 14(1):14254
40. Rizwan M, Shoukat A, Ayub A, Razzaq B, Tahir MB (2021) Types and classification of nanomaterials. In: Nanomaterials: synthesis, characterization, hazards and safety. Elsevier, pp 31–54
41. Samarajeewa AD, Velicogna JR, Schwertfeger DM, Princz JI, Subasinghe RM, Scroggins RP, Beaudette LA (2021) Ecotoxicological effects of copper oxide nanoparticles (nCuO) on the soil microbial community in a biosolids-amended soil. Sci Total Environ 763:143037
42. Shao S, Wu W, Meng F, Duan M, Zhang M (2023) Copper-based nanomaterials: a rising star in cancer therapy. Adv Therapeutics 6(10):2300049
43. Simamora AJ, Hsiung TL, Chang FC, Yang TC, Liao CY, Wang HP (2012) Photocatalytic splitting of seawater and degradation of methylene blue on CuO/nano TiO_2. Int J Hydrogen Energy 37(18):13855–13858
44. Sreeju N, Rufus A, Philip D (2016) Microwave-assisted rapid synthesis of copper nanoparticles with exceptional stability and their multifaceted applications. J Mol Liq 221:1008–1021
45. Su C, Chen A, Liang W, Xie W, Xu X, Zhan X, Peng C (2024) Copper-based nanomaterials: opportunities for sustainable agriculture. Sci Total Environ 171948
46. Sutulienė R, Ragelienė L, Duchovskis P, Miliauskienė J (2022) The effects of nano-copper,-molybdenum,-boron, and-silica on pea (Pisum sativum L.) growth, antioxidant properties, and mineral uptake. J Soil Sci Plant Nutrit 1–14
47. Švarcová S, Hradil D, Hradilová J, Čermáková Z (2021) Pigments—copper-based greens and blues. Archaeol Anthropol Sci 13:1–29
48. Tripathi S, Mahra S, Tiwari K, Rana S, Tripathi DK, Sharma S, Sahi S (2023) Recent advances and perspectives of nanomaterials in agricultural management and associated environmental risk: a review. Nanomaterials 13(10):1604
49. Wahab A, Muhammad M, Ullah S, Abdi G, Shah GM, Zaman W, Ayaz A (2024) Agriculture and environmental management through nanotechnology: eco-friendly nanomaterial synthesis for soil-plant systems, food safety, and sustainability. Sci Total Environ 171862
50. Wang S, Fu Y, Zheng S, Xu Y, Sun Y (2022) Phytotoxicity and accumulation of copper-based nanoparticles in brassica under cadmium stress. Nanomaterials 12(9):1497
51. Yusefi-Tanha E, Fallah S, Rostamnejadi A, Pokhrel LR (2020) Root system architecture, copper uptake and tissue distribution in soybean (Glycine max (L.) Merr.) grown in copper oxide nanoparticle (CuONP)-amended soil and implications for human nutrition. Plants 9(10):1326

Toxicity and Safety Concerns of Copper-Based Nanomaterials in Agriculture

Zohreh Shirkhani, Ahmad Mirjalili, and Ali Lafta

Abstract Copper-based nanomaterials (Cu-NMs) have been used increasingly in agriculture because they can increase crop productivity, pest control, and soil health. However, their toxicity and environmental safety have been a concern. Because of their small size and large surface area, Cu-NMs can dissolve with the release of Cu ions to disrupt biological processes. This can result in oxidative stress and plant health and productivity are dis-compromised. Additionally, Cu-NMs may inhibit nutrient uptake resulting in nutrient imbalances and lowered plant growth. Finally, Cu-NMs can have significant environmental impacts due to persistent occurrence in soil and water systems and the ability to impact soil microbiota and non-target organisms. There is a concern about human health risks, including heavy metal exposure from Cu-NMs that are bioaccumulating in food chains. Safe application practices, use of biodegradable or less toxic nanomaterial formulations and optimizing nanoparticle (NPs) concentrations are critical so that Cu-NMs can relieve toxicity in agricultural systems. An essential aspect of developing safer, more effective agricultural applications based on Cu-NM is that advanced nanomaterials with controlled release properties reduce the probability of Cu-NM accumulating in plants and soil. Finally, while they offer great promise for increasing agricultural productivity and crop protection, Cu-NMs can end up being toxic or pose unacceptable safety concerns. This chapter aims to provide overview on toxicity and safety concerns of copper-based nanomaterials in agriculture.

Keywords Copper toxicity · Nanomaterial · Oxidative stress · Crop performance

Z. Shirkhani (✉)
Department of Plant Sciences, Faculty of Biological Sciences, Kharazmi University, Tehran, Iran
e-mail: z.shirkhani@khu.ac.ir

A. Mirjalili
Department of Horticultural Science and Agronomy, Science and Research Branch, Islamic Azad University, Tehran, Iran

A. Lafta
Department Biology, Razi University, Kermanshah, Iran

R. K. Bachheti et al. (eds.), *Copper-Based Nanomaterial for Agricultural Practices*, Smart Nanomaterials Technology, https://doi.org/10.1007/978-981-96-8610-0_3

1 Introduction

Nanomaterials (NMs) are materials whose properties differ from bulk materials in terms of size, processibility, mechanical and chemical stability, dispersion stability, and biological specificity at the nanoscale. With applications in a variety of scientific and technological fields of interest, they have attracted considerable research interest. Copper based nanomaterials (Cu-NMs) are of interest because they exhibit antimicrobial properties, nutrient delivery capabilities, and the ability to improve crop protection [8, 38, 78, 86]. Cu-nMs are formulated as nanoscale products and can control plant diseases with targeted and effective control in diverse agricultural uses, such as pesticides, fungicides, and fertilizers. Despite their safety and geoenvironmental concerns, they have raised concerns about their toxicity to plants, soil microbiota, aquatic systems, and contaminated human health through food consumption [36]. Toxicity depends on particle size, shape, surface charge, concentration, and environmental context. Smaller NPs have greater reactivity and greater potential for harm. Transformations of Cu nanomaterials in the soil can release free Cu ions, which are more toxic than bulk Cu itself to organisms. Development of safe and sustainable agricultural practices requires an understanding of these interactions. Although regulatory bodies are attempting to assess the risks posed by nanomaterials in agriculture, the lack of comprehensive study of long-term exposure and environmental persistence means safety assessments still need to be completed. However, the agricultural industry needs to work on maximizing benefits and reducing the risk [6, 8, 68, 70].

2 Copper Toxicity

Cu is an essential micronutrient, playing a major role in the biological processes of photosynthesis and enzyme function. However, large excesses of Cu can be toxic for plants and the environment, especially in agricultural settings where Cu based products, including Cu NPs, have been used extensively as pesticides, fungicides and fertilizers. The recent increase in the use of Cu-NMs has brought the focus to Cu toxicity, which is owing to high reactivity and potential for environmental accumulation. Physiological and biochemical disruptions are induced by Cu toxicity in plants, including chlorosis, stunted growth, leaf necrosis and reduced nutrient uptake. The tolerance of Cu and the accumulation of Cu in the plant tissue can threaten human and animal health if consumed [75]. Gastrointestinal, liver and kidney damage and neurological effects are human health concerns related to Cu toxicity. To mitigate Cu cytotoxicity, cu based nanomaterials must be administered with optimum Cu dosages, substituted with other materials and monitored for Cu levels in soil and water [3, 56].

2.1 *Toxicity and Safety of Copper Nanomaterials in Air*

Owing to their widespread use in many industries, including agriculture, electronics, and medicine, Cu-NMs are actively studied. However, their release into the environment—and especially, into the air—are of concern with regard to their toxicity and health risks. Cu-NMs have a high surface area, which increases their reactivity and small size, facilitating their possibility of toxicity after inhalation or contact with living organisms [75].

Cu-NMs are airborne and dispersed in production, handling, or application processes, particularly in industries or agriculture. They are small enough to stay suspended in the air for extended periods before being inhaled by humans and animals. These NPs can be inhaled once and go deep into the respiratory system to the alveoli, which may be absorbed into the bloodstream and transported to other organs [20].

Airborne Cu-NMs can have toxicological effects on respiratory, systemic, and environmental effects. ROS generation can induce oxidative stress, inflammation, and cytotoxicity, resulting in cellular damage, inflammation, and impaired lung function. Liver and kidney damage, neurotoxic effects, and cardiovascular effects are systemic effects. Environmental concerns related to the release of airborne Cu-NMs into the air include soil contamination, aquatic toxicity, and bioaccumulation in plants and animals. Thus, strict workplace safety measures are needed to prevent the release of Cu-NMs into the air [75].

2.2 *Toxicity and Safety of Copper Nanomaterials in Water*

Concern has arisen over Cu-NMs being released into aquatic environments due to the increasing use of Cu-NMs in agriculture, medicine, and manufacturing industries. These materials have known antimicrobial and catalytic properties that can find their way into bodies of water through runoff and wastewater discharges or as much as accidental spills, and they could pose threats to aquatic animals and their environments. The toxicity and safety of Cu-NMs in water are important issues to be understood to make possible environmental sustainability and prevent the harmful impact of Cu-NMs on aquatic life and human health. Cu nanomaterials are transported to water bodies through agricultural runoff, industrial discharge, or wastewater treatment [15]. Aquatic environments can be highly reactive and cause a toxicological impact on aquatic organisms such as fish, algae, invertebrates, and microorganisms. They impact oxidative stress, gill dysfunction, respiratory impairment, and fish asphyxiation. In addition to preventing photosynthesis, Cu poisoning can also prevent algae and plants from absorbing nutrients. Cu-NMs have the potential to bioaccumulate since they are very poisonous to snails, mussels, and daphnia. Particle size and shape, water chemistry, and aggregation all influence the environmental destiny of Cu-NMs in water. Regulatory standards and guidelines can effectively ensure the safety and health of aquatic organisms and their ingestion [44, 96]. Their antibacterial properties have been helpful in their increasing use in agriculture, and therefore, they have a potential role as a fertilizer, disease control, and growth promoters. Questions about their toxicity and potential to damage the environment, soil health, and food safety are also raised when they are added to the soil. Cu-NMs can enter the soil through industrial runoff treated wastewater biosolids or agricultural usage [34].

2.3 Toxicity and Safety of Copper Nanomaterials in Soil

These toxicological effects on soil ecosystems of Cu-NMs include the disruption of soil microorganisms, reduced enzyme activity, and soil fauna. Other important contributors to Earth's soil structure and plant health are earthworms, nematodes, and arthropods. Cu-NMs can cause oxidative stress in earthworms, as evidenced by cell damage and inhibition of earthworms in the contaminated soil [84].

Cu-NMs at high soil concentrations could cause phytotoxicity, oxidative stress, diminished growth, and damaged roots. Food grown on land that is corrupted with the heavy metal copper suffers from nutrient deficiencies (in the broadest sense), caused most importantly by burning of iron and zinc food. Such inhibition of root growth may lead to poor plant health and reduced crop yields [68, 70, 72].

Although the factors controlling the environmental fate of Cu NMs in soil are still uncertain, particle size and surface properties are known to affect particle behavior. In soil, smaller NPs have greater surface areas and are more reactive and mobile, able to get into the soil more than larger ones. To ensure safe use of Cu-NMs in agriculture with little or no unwanted effects, it is imperative to determine how they interact with soil ecosystems [84] (Fig. 1).

3 Copper Threshold in the Environment

The threshold levels of Cu NPs in soil, water, air, and plants can vary based on factors such as particle size, exposure duration, environmental conditions, and the biological species affected. General guidelines and thresholds for Cu concentrations have been established to limit toxicity.

Although total Cu content is usually limited because of its potential toxicity to plants, microbes, and soil-dwelling species, soil thresholds are less well-defined than those for bulk Cu. The usual range of allowable total Cu content for agricultural soils is 20–100 mg/kg of dry soil, contingent upon soil type and regulatory requirements. Cu concentrations over 100 mg/kg have the potential to be harmful to delicate soil organisms and to suppress microbial activity, which is essential for the cycling of nutrients. Higher amounts may potentially influence plant growth and root development. NPs Cu can behave differently than bulk Cu, making specific thresholds for Cu NPs more complex to establish.

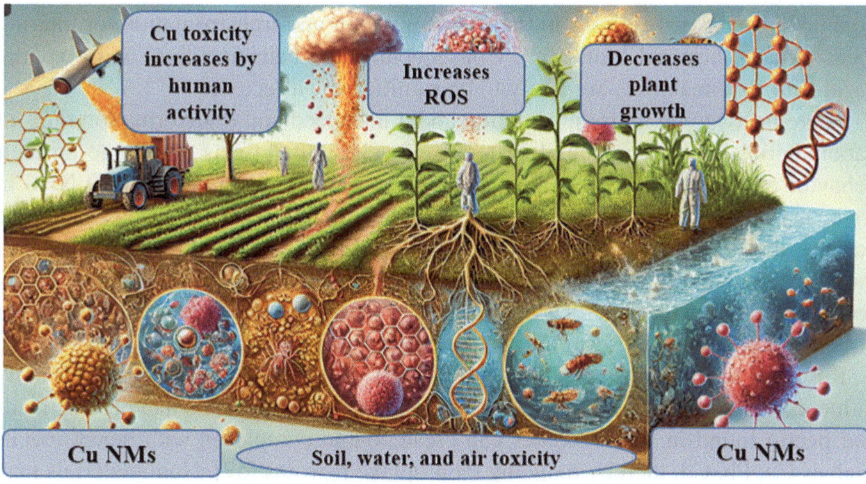

Fig. 1 The toxicity of copper nanomaterials on plant growth and development

Occupational exposure limits, as opposed to environmental norms, primarily set water thresholds. In industrial environments, breathing in Cu NPs can harm one's health [81]. PELs for respirable Cu dust and mists (including NPs) in the air are set by the U.S. Occupational Safety and Health Administration (OSHA) at 0.1 mg/ m^3 and 1 mg/m^3, respectively, for Cu fumes and dust. Since NPs can be hazardous when breathed over extended periods of time, no defined environmental criteria for Cu NPs in ambient air have been determined. However, it is important to monitor particulate matter (PM) levels in urban or industrial regions [54].

The types of plants, the characteristics of the soil, and the exposure circumstances all affect plant thresholds. Plants require extremely little amounts of Cu (5–30 mg/ kg of dry plant tissue is the ideal range). Depending on the species and soil type, concentrations of Cu in plant tissues over 20–100 mg/kg render most crops poisonous. Because Cu NPs are smaller than bulk Cu, they may be absorbed by plant roots more readily, which may cause toxicity in lower quantities [2].

4 Uptake and Transport of Nanomaterials in Plants from the Rhizosphere

Nanomaterials are very dynamic and reactive due to their unique physicochemical properties compared to their bulk counterparts. Typically, metal-based NPs like zinc oxide, Cu oxide, and silver can engage in chemical reactions with metal ions, minerals (such as sulfide and phosphate), or organic compounds that are widely present in the environment and living organisms, leading to their dissolution and release [22]. Nanomaterials can also physically interact with inorganic ions, biomolecules, and natural organic substances, resulting in their accumulation and transformation, or altering their surface chemical traits [65]. Consequently, the behavior, fate, and potential hazards of nanomaterials are likely to be significantly affected.

There needs to be more information about the speed and transformation rates of nanomaterials in environmental systems, as well as the effects of change on their behavior and toxicity levels. For instance, zinc oxide nanoparticles (ZnO NPs) in soybean roots are transformed into Zn^{+2}, $ZnNO_3$, or zinc acetate, while cerium oxide nanoparticles (CeO NP) remain unaffected [48]. Inside the plant, NPs may undergo modifications facilitated by various plant components. Plant roots can release many secretions, including mineral ions, small organic molecules, and macromolecular organic compounds that form the rhizosphere around the root [7].

Research has demonstrated that root secretions in the rhizosphere significantly influence how plants interact with their environment. Compounds like organic acids and pectin found in root exudates can create stable complexes with heavy metals such as Cu, Pb, Cd, thereby reducing their uptake by roots [57]. Nanomaterials may also undergo physicochemical changes in interaction with root secretions because nanomaterials frequently interact directly with plant roots. These transformations affect the final fate and toxicity of nanomaterials in plants. Generally, nanomaterials can be readily accumulated and absorbed on the roots' surfaces [52].

NPs initially interact with plant roots through adsorption on the root surface, where root hairs, which carry a negative charge, attract positively charged NPs [49]. As lateral roots grow, they provide more surfaces for these NPs to attach, aiding their movement into the root system [62]. The root epidermis, which has a structure and role similar to that of leaf surfaces, contains regions that are not fully developed, permitting NPs to make direct contact and penetrate [68, 70]. The root cell wall of the semi-permeable quality with small pores prohibits the passage of larger particles [63]. Ali et al. [4] have indicated that some NPs can disrupt the plasma membrane and form new pores for larger particles. Ionic transport, endocytosis, and interactions with proteins on the cell membrane are some routes for plant cell uptake of NPs [49]. Hydrophilic pathways can help with some, while the small pore sizes greatly hinder their utility. The uptake of NPs relies on endocytosis as a mechanism to pass particles $< 1\ \mu m$ in size through the cell membrane irrespective of particle size [51]. In addition, NPs can adjoin to transportation proteins on the external epidermis to catalyze their ingestion [30].

After NPs interact with plant roots, they are taken up by root hair cells and can pass through the cell wall to enter the endothelial layer through either the symplast or apoplast pathways. NPs are then moved up to aboveground parts of the plant by xylem vessels [83].

5 Uptake and Transport of Nanoparticles in Plant Leaves

In agricultural applications, NPs are sprayed on the leaf surface and absorbed in the plant through the cuticle or stomata. It also takes the role of a shield against moisture loss and the entry of NPs [63]. It has two channels: Its diameters vary from 0.6 to 4.8 nm, and it is hydrophilic and lipophilic [23]. Hydrophilic channels facilitate the uptake of NPs < 4.8 nm, and lipophilic channels amplify the uptake of dansyl-lipophilic NPs [9]. It has been found that NPs can accumulate in the leaf epidermis and vascular tissue. The scientists also argued that NPs can pass through the stomata. The exchange of water and gases through stomata, which are responsible for the management of this process, ranges from 10 to 100 μm in size and have a density dependent on plant species [32], but the exact size exclusion limit for the diffusion of NPs through stomata remains unknown [32].

It has been demonstrated that NPs intake by leaves of plants can be transported to many parts of plant body, including stems, roots, flowers, fruits, and even the rhizosphere soil [80, 87, 88]. Following the spraying of CuO-NPs on maize leaves, the CuO-NPs were found to be present on the surface of leaves and could transfer through plasmodesmata [1]. For another investigation, gold NPs deposited on watermelon leaf surfaces were found to move into the leaf cells, and could be tracked to stems and roots [71]. In addition, for 72 h cerium oxide (CeO_2) NPs migrate to cucumber roots from cucumber leaves after exposure [33].

6 Toxicity Effects of Copper Nanomaterials on Plants

Cu NMs cause large changes in root length, shoot height, leaf size, and major perturbations in plant physiology. Finally, they can disturb photosynthesis, nutrient uptake, and antioxidant defense mechanisms and influence plant growth and development, paving the way to identifying risks that may threaten agriculture.

6.1 Copper Toxicity on Plant Morphological Parameters

Cu is a key proteins cofactor [14]. Optimal Cu amounts as a trace element are necessary for cellular activities but excess amounts are detrimental to plant development and survival [64]. As an integral part of several plant metabolic functions, Cu affects growth and productivity tremendously when roots take up too much Cu. These changes arise due to root modifications, nutritional disorders, oxidative stress, and increased respiration of ROS [19, 53]. Cu stress evidences biomass reduction, arrested root growth, chlorosis, bronzing, and necrosis resulting from increased ROS generation and deleterious cell interactions [42].

High concentrations of Cu effect on the plant's growth. When plants experience stress from Cu, their root structures undergo alterations [13], since the roots are the first to come into contact with the pollutant. As Cu levels rise, there is a reduction in the biomass of both the roots and the shoots [25]. Cu toxicity manifests through various issues, such as harm to the protective root layer, reduced growth of root hairs, and notable changes in the overall root architecture [77, 89].

Cu occurs in the roots and can move into the shoots [92]. At elevated levels, Cu present in large amounts in the soil can be harmful and inhibit plant growth as Cu has toxic effects on plants [2, 18, 21]. Growth and stability during later development stages are both hindered by high Cu. Like reductions in growth, the observed growth reduction also comes from a decreased ability to assimilate carbon [14, 43].

Additionally, an area in the roots is responsible for how much the plant can absorb nutrients. As the number of roots begins to grow, so does the plant's capacity to absorb these necessary elements, which, in turn, will benefit the plant overall. However, as Cu exposure reduces root volume, the major loss of root biomass is associated with the process of cell division [12, 13], and this reduction in root volume is reflected in opposite measurements. According to Batool et al. [11], root development significantly declines in plants under high metal stress, and thicker root cell walls result from a commensurate drop in cell division. According to Chen et al. [16] and Jung et al. [39], common indications of sedimentary Cu toxicity include slowed root growth, leaf chlorosis, withering leaves, necrosis, and root abnormalities. The xylem with high Cu levels may cause the cell walls to thicken and harden by reducing the leaf area with high Cu levels. Therefore, these ultimately restrict the flexibility of cells and the formation of leaves [27]. Xuerong et al. [91] found that CuO NPs increased the growth of Brassica chinensis L. by increasing biomass, net photosynthesis, root length, number of root tips, and concentrations of mineral elements. However, Cu NPs and $CuSO_4$ have negatively affected plant development with decreases in biomass, net photosynthesis, root length, and content of mineral elements. Exposure to CuO NPs was generally favorable for plant growth. The results help further understand the effects on B. chinensis of various forms of copper phytotoxicity. Thus, *e* suggest that CuO NPs can be used to improve nutrition and enhance the growth of edible plants. In the rice Nipponbare variety, CuO NPs' introduction increased lipid peroxidation in the root and shoot tissues. It was found that ethylene influenced this reaction to oxidative stress [5]. CuO NPs caused damage to the cell walls of roots, and that damage adversely affected root development by reducing the expression of genes that control cell wall organization [59]. The results of the study show that applying Cu will lessen growth characteristics. Furthermore, symptoms of distress, including chlorosis, leaf purpling, and necrosis, were observed upon administration of Cu stress.

6.2 Effect of Copper Toxicity on Physiological Parameters

Heavy metal pollution induces many harmful physiological and biochemical alterations in plants. Thus, developing and using various approaches to help these organisms mitigate the adverse impacts of heavy metal exposure is significant [66, 67]. Cu excess compromises the structural stability of chloroplasts and thylakoid membranes, leading to oxidative stress in plant cells. This stress not only inhibits the necessary electron transfer necessary for photosynthesis but also lowers the quantities of key photosynthetic pigments and electron transporters [29]. According to the study's findings, Cu mostly interferes with Crassula helmsii's photosynthesis at the reaction center of photosystem II (PSII). Slow chlorophyll incorporation into the photosystems and pigment buildup can both contribute to Cu-related chlorosis [41].

Cu stress altered photosynthetic pigment concentrations in Lens culinaris Medik cv. BARI Lentil-7 [35]. Their study also showed that excessive application of Cu at 3 mM decreased the amounts of these pigments. As also noted in [11], the loss in photosynthetic pigments may impair the plants' capacity to carry out photosynthesis in a way that hinders their growth. Chloroplast size and quantity are reduced when Origanum vulgare is exposed to Cu at concentrations between 17 and 25.5 mM [61]. Furthermore, they discovered that Cu stress caused grana thylakoids to depolarize, degrade, and swell, respectively. Cu exposure frequently causes globulin to accumulate more in plastids, which may indicate a change in the structure and function of cells.

In the photosynthetic electron transport chain, Cu is used by photosynthetic system II to catalyze electron transport. However, excessive amounts can impede this process. For instance, elevated Cu causes a decrease in chlorophyll, which lowers photosynthetic efficiency and impairs photosystem II activity in *Hippophae rhamnoides* [14]. Increased Cu concentrations are likewise harmful to *Synechocystis*sp. thylakoid membranes, just like they are to *Audouinella*spp. [90]. Lower eggplant development is the result of CuO NP treatments. DNA damage, reduced total chlorophyll, and oxidative stress have all been connected to this decline [10]. For example, when rice shoots were exposed to 250 mg L^{-1} CuO NPs, their electrical conductivity and malondialdehyde (MDA) levels increased by 67% and 83%, respectively. On the other hand, at this concentration, the levels of SOD and CAT in rice leaves dropped. However, when rice roots were exposed to 125 mg/L CuO NPs, their SOD activity increased 1.66 times. Also, the amount of photosynthetic pigments in the leaves of rice plants decreased when CuO NPs were applied [93]. A particular plant species, its genetic variety, the concentration of CuO NPs, and the application technique all impact how CuO NPs affect plants [24]. Consequently, regulating Cu levels in plants is essential to escape from the detrimental effects of photosynthesis (Table 1).

Table 1 Toxicity of copper nanomaterials on plant growth and development

Author	Plant	Findings
Mottaleb and Hassan [58]	*Allium cepa*	Using newly reclaimed sandy soils, the researchers found that CuO NPs treatment significantly enhanced growth parameters, including plant height, leaf number, fresh and dry weight, yield, and bulb quality. Despite being the cheapest form of Cu supplementation, CuO NPs was found to be safe and led to the highest onion yield and quality
Feil et al. [26]	*Cucumis sativus*	The use of cupric fungicides in crop-defense programs has led to the accumulation of Cu in soil, potentially causing toxicity in plants. This toxicity can lead to nutritional imbalances and constraints on plant growth, particularly in relation to phosphorus (P), an essential plant macronutrient. Studies on hydroponically grown cucumber plants showed that high Cu concentrations stunted shoot and root growth, diminished P influx rate, and upregulated two P transporter genes
Da Costa et al. [17]	*Oryza sativa*	CuO NPs caused growth decline, photo-phosphorylation, and carbon dioxide assimilation, while bulk Cu caused oxidative stress. NPs accumulated in chloroplasts, causing damage to thylakoid membranes without oxidative damage
Zhang et al. [95]	*Triticum aestivum*	Exposure to 15.6 μM nCu reduced root growth rate to 60% and stimulated lateral root formation. The expression of genes involved in cell proliferation and programmed cell death decreased in Cu^{2+}- plants, while only one gene was down-regulated in nCu-treated plants. Oxidative stress and root growth inhibition were the main causes of nCu toxicity, with nCu induced antioxidant synthesis to alleviate phytotoxicity
Rajput et al. [69]	*Hordeum sativum*	CuO NPs inhibit growth by affecting germination rate, root and shoot lengths, photosystem II maximal quantum yield, and transpiration rate. The study also found that the Cu content of roots and leaves was 5.7 and 6.4-fold higher in CuO NPs-treated plants compared to the control
Liu et al. [45]	*Oryza sativa*	The increasing release of Cu oxide nanoparticles into the environment may pose risks to rice plants, reducing their quality and quantity in the human diet. Arsenic contamination in rice has decreased yield and raised health concerns. Few studies have addressed CuO NP phytotoxicity to rice and the interactions between CuO NPs and As

7 Detoxification of Copper-Based Nanomaterials

Cu-NMs' conductive, catalytic, and antibacterial qualities have made them extensively used in a variety of sectors. But as they are used more, worries about their biological and environmental toxicity have grown. Cu-NMs can produce ROS and release Cu^{2+}, all of which are linked to oxidative stress, cell injury, and ecological devastation. Therefore, detoxifying Cu-NMs is crucial to reducing their harmful effects on the environment and biological systems [79].

7.1 Mechanisms of Copper Detoxification in Plants

In plants, the mechanism of removing Cu toxicity occurs by induction of expression of functional genes and antioxidants [40]. ROS are produced under Cu sulfate stress in plants and there are three mechanisms to balance the level of ROS: (i) Chelation or precipitation of Cu ions by plant secretions to reduce or prevent Cu absorption by roots; (ii) Inducing genes to reduce Cu ion concentration as well as Cu ion induced ROS; (iii) Removal of ROS by various antioxidants to reduce Cu ion induced toxic effects [94].

Root secretory substances are substances that are secreted from the root system to the environment during the plant growth process and have the ability to change the physical, chemical, and biological properties of the rhizosphere and improve plant growth. Substances secreted from roots are divided into three categories: (1) substances secreted from root cells, such as sugars, amino acids, and vitamins; (2) metabolites actively secreted by root cells, such as enzymes, hormones, phenols, and organic acids; (3) The materials that are created by the decomposition of bacteria after the dumping of plant residues [47].

Lyubenova et al. [50] observed that oxalic acid, malic acid, static acid, and tartaric acid are secreted from the roots of *Typha radix* L. under Cu stress. Heavy metals can also form stable compounds with amino, carboxyl, and hydroxyl functional groups of histidine and proline, detoxifying them [37].

Relocating Cu ions to vacuoles or other organelles helps plants tolerate Cu stress. Cu is found in the soil in terms of compounds known as Cu minerals, but these decompose and break down into Cu^{+} and Cu^{2+} ions. Cu^{2+} ions are transformed to Cu^{+} by ferric reductase oxidase 4/5 in the rhizosphere. COPT1/2 and ZIP2/4 transporters transport Cu^{+} and Cu^{2+} into root cells, subsequently transporting these ions to mesophyll cells. Moreover, Cu ions can be directly transported from the surrounding environment to the leaf sheath cells via the HMA and COPT [55].

Plastid chaperone 1 (PCH1) and CCS are crucial in the distribution of Cu^{+} ions (within chloroplasts. PCH1 facilitates the movement of Cu^{+} from the cytoplasm to the chloroplast membrane, subsequently delivering Cu^{+} to the Cu transporter HMA6 found in the inner membrane. Ultimately, HMA8 supplies Cu to plastocyanin within the thylakoid lumen [64, 85].

Under Cu stress, antioxidant enzymes are critical for ROS scavenging [28]. Activation or inactivation of antioxidant enzymes to respond to Cu-induced oxidative stress depends on ROS and plant species [76]. It was observed that proline content and peroxidase (POX) activity increased significantly in Barley after three days of treatment with 50 μM $CuCl_2$. In rice seedlings, *superoxide dismutase* activity increased and catalase (CAT) activity decreased after treatment with 200 μM Cu [82]. Saleem et al. [74] found that with the increase of Cu concentration in the soil, the activity of SOD and POX in *Linum usitatissimum* L. showed an upward trend and the enzyme activity reached its highest level after treatment with 600 mg/kg of Cu.

Glutathione (GSH) is an essential tripeptide that functions as a metal-binding agent and antioxidant, playing a critical role in managing the cell cycle, detoxifying harmful substances, and bolstering antioxidant defenses. It improves the tolerance to Cu through the ascorbic acid glutathione cycle [60]. Phytochelatins (PCs) are important metal-binding agents that aid in regulating the balance of Cu ions. These compounds are produced from GSH and can create harmless complexes with free metal ions, thus averting cellular damage [73]. Metallothioneins (MTs) are proteins rich in cysteine that bind various metal ions, including Cu, and help maintain Cu balance within the cytoplasm [76]. Research by Liu et al. [46] demonstrated that increasing the expression of the rice gene OsMT2c in Arabidopsis enhanced its capacity to eliminate ROS and improved its tolerance to Cu. Exposure to Cu prompts the activation of different MT genes in a manner that depends on the concentration.

Cu-NMs typically cause harm by one of two methods including ion release: $Cu2^{+}$, which interacts with biomolecules and causes cellular damage, is released during the breakdown of Cu-NMs. ROS formation: Cu-NMs accelerate the production of ROS, which results in oxidative stress and damages proteins, lipids, and DNA. Overproduction of ROS is associated with apoptosis, inflammation, and defective mitochondria [79].

8 Environmental and Human Health Implications

Cu-NMs are widely used in industries like electronics, agriculture, and medicine, but their increasing production raises concerns about their environmental and human health impacts. Cu-NMs can accumulate in soil, water, and air, disrupting ecosystems and affecting plant growth and metabolism. They can also be transported over long distances by wind or water, spreading contamination to previously unaffected areas. Human health impacts include oxidative stress, inflammation, and potential damage to vital organs. Chronic exposure may increase the risk of respiratory issues, cardiovascular diseases, and neurotoxicity due to their ability to cross biological barriers. The long-term health effects of Cu-NMs are not yet fully understood, but prolonged exposure could result in cumulative toxicity and adverse health outcomes. Therefore, while Cu-NMs offer technological advantages, their environmental and health risks must be managed carefully, with regulatory frameworks, risk assessments, and safer design strategies being crucial [31].

9 Future Perspective

Because of their antibacterial qualities, capacity to promote plant development, and function in nutrient delivery systems, Cu-NMs have the potential to increase agricultural output. However, to guarantee their safe and sustainable use in agriculture, issues with their toxicity and safety require a forward-thinking strategy. Some future directions include the creation of ecologically acceptable Cu-NMs, thorough toxicity analyses, and the establishment of safety regulations and standards.

By including Cu-NMs in precision farming and other sustainable farming methods, Cu-NMs can increase crop output while reducing negative side effects. For Cu-NMs to be used successfully and safely, farmer education and public perception are essential. Building trust and ensuring responsible use at the farm level can be achieved through open and honest discussion about possible hazards and safety precautions.

10 Conclusions

Owing to their highly effective antibacterial properties, Cu-NMs hold promise for use in agriculture. They act and penetrate microorganism cell membranes, which results in better results even when a smaller amount is used. Apart from being used as vectors for pesticides and herbicides through which they improve performance and minimize effects on the ecosystem, Cu-NMs can be used to improve the uptake of Cu by plants, which is vital in photosynthesis and enzyme function. They can also be used in seed coatings to guard the seed against diseases that are conveyed by the soil at initial growth. However, the following are safety concerns: They have limited size, contact with biological systems, water and soil pollution, accumulation in food chains, and adverse effects on human health. Designing safer Cu-NMs, long-term environmental monitoring, and robust regulatory frameworks are research areas for safe integrating Cu-NMs into agricultural practices. It presents environmental and health hazards, is a resource for developing safer nanomaterial applications, sustainable practices, and regulatory decisions, and is an important reference for nanotechnology and environmental science studies.

Competing Interests The authors declare that they have no competing interest.

Declaration of Generative AI and AI-Assisted Technologies in the Writing Process During the preparation of this work the book chapter author(s) have not taken any assistance from AI, and the book chapter author(s) reviewed and edited the content as needed and take(s) full responsibility for the content of the publication. All authors read and approved the final version of the manuscript.

References

1. Adhikari T, Kundu S, Biswas AK, Tarafdar JC, Subba Rao A (2015) Characterization of zinc oxide nano particles and their effect on growth of maize (*Zea mays* L.) plant. J Plant Nutr 38(10):1505–1515
2. Adrees M, Ali S, Rizwan M, Ibrahim M, Abbas F, Farid M, Zia-ur-Rehman M, Irshad MK, Bharwana SA (2015) The effect of excess copper on growth and physiology of important food crops: a review. Environ Sci Pollut Res22:8148–8162

3. Alengebawy A, Abdelkhalek ST, Qureshi SR, Wang MQ (2021) Heavy metals and pesticides toxicity in agricultural soil and plants: ecological risks and human health implications. Toxics 9(3):42
4. Ali S, Mehmood A, Khan N (2021) Uptake, translocation, and consequences of nanomaterials on plant growth and stress adaptation. J Nanomater 2021(1):6677616
5. Azhar W, Khan AR, Salam A, Ulhassan Z, Qi J, Shah G, Liu Y, Chunyan Y, Yang S, Gan Y (2023) Ethylene accelerates copper oxide nanoparticle-induced toxicity at physiological, biochemical, and ultrastructural levels in rice seedlings. Environ Sci Pollut Res 30(10):26137–26149
6. Bachheti RK, Archana Bachheti AH (eds) (2024) Metal and metal-oxide based nanomaterials. In: Synthesis, agricultural, biomedical and environmental interventions. Springer, Singapore. https://doi.org/10.1007/978-981-99-7673-7
7. Bais HP, Weir TL, Perry LG, Gilroy S, Vivanco JM (2006) The role of root exudates in rhizosphere interactions with plants and other organisms. Annu Rev Plant Biol 57(1):233–266
8. Bakshi M, Kumar A (2021) Copper-based nanoparticles in the soil-plant environment: assessing their applications, interactions, fate and toxicity. Chemosphere 281:130940
9. Banerjee K, Pramanik P, Maity A, Joshi DC, Wani SH, Krishnan P (2019) Methods of using nanomaterials to plant systems and their delivery to plants (mode of entry, uptake, translocation, accumulation, biotransformation and barriers). In: Advances in phytonanotechnology. Academic Press, pp 123–152
10. Baskar V, Nayeem S, Kuppuraj SP, Muthu T, Ramalingam S (2018) Assessment of the effects of metal oxide nanoparticles on the growth, physiology and metabolic responses in in vitro grown eggplant (*Solanum melongena*). 3 Biotech 8:1–12
11. Batool R, Hameed M, Ashraf M, Ahmad MSA, Fatima S (2015) Physio-anatomical responses of plants to heavy metals. In: Phytoremediation for green energy, pp 79–96
12. Bochicchio R, Sofo A, Terzano R, Gattullo CE, Amato M, Scopa A (2015) Root architecture and morphometric analysis of *Arabidopsis thaliana* grown in Cd/Cu/Zn-gradient agar dishes: a new screening technique for studying plant response to metals. Plant Physiol Biochem 91:20–27
13. Cai S, Xiong Z, Li L, Li M, Zhang L, Liu C, Xu Z (2014) Differential responses of root growth, acid invertase activity and transcript level to copper stress in two contrasting populations of *Elsholtzia haichowensis*. Ecotoxicology 23:76–91
14. Cambrollé J, García JL, Figueroa ME, Cantos M (2015) Evaluating wild grapevine tolerance to copper toxicity. Chemosphere 120:171–178
15. Cao X, Yu ZX, Xie M, Pan K, Tan QG (2023) Higher risks of copper toxicity in turbid waters: quantifying the bioavailability of particle-bound metals to set site-specific water quality criteria. Environ Sci Technol 57(2):1060–1070
16. Chen J, Shafi M, Li S, Wang Y, Wu J, Ye Z, Peng D, Yan W, Liu D (2015) Copper induced oxidative stresses, antioxidant responses and phytoremediation potential of Moso bamboo (*Phyllostachys pubescens*). Sci Rep 5(1):13554
17. Da Costa MVJ, Kevat N, Sharma PK (2020) Copper oxide nanoparticle and copper (II) ion exposure in Oryza sativa reveals two different mechanisms of toxicity. Water Air Soil Pollut 231:1–16
18. DalCorso G, Manara A, Piasentin S, Furini A (2014) Nutrient metal elements in plants. Metallomics 6(10):1770–1788
19. De Conti L, Cesco S, Mimmo T, Pii Y, Valentinuzzi F, Melo GW, Ceretta CA, Trentin E, Marques AC, Brunetto G (2020) Iron fertilization to enhance tolerance mechanisms to copper toxicity of ryegrass plants used as cover crop in vineyards. Chemosphere 243:125298
20. De la Campa AMS, Sánchez-Rodas D, Castanedo YG, de la Rosa JD (2015) Geochemical anomalies of toxic elements and arsenic speciation in airborne particles from Cu mining and smelting activities: influence on air quality. J Hazard Mater 291:18–27
21. De Marco R, da Silva RF, Andreazza R, Da Ros CO, Scheid DL, Bertollo GM (2016) Copper phytoaccumulation and tolerance by seedlings of native Brazilian trees. Environ Eng Sci 33(3):176–184

22. Dimkpa CO, Latta DE, McLean JE, Britt DW, Boyanov MI, Anderson AJ (2013) Fate of CuO and ZnO nano-and microparticles in the plant environment. Environ Sci Technol 47(9):4734–4742
23. Eichert T, Kurtz A, Steiner U, Goldbach HE (2008) Size exclusion limits and lateral heterogeneity of the stomatal foliar uptake pathway for aqueous solutes and water-suspended nanoparticles. Physiol Plant 134(1):151–160
24. Feigl G (2023) The impact of copper oxide nanoparticles on plant growth: a comprehensive review. J Plant Interact 18(1):2243098
25. Feigl G, Kumar D, Lehotai N, Tugyi N, Molnár Á, Ördög A, Szepesi Á, Gémes K, Laskay G, Erdei L, Kolbert Z (2013) Physiological and morphological responses of the root system of Indian mustard (*Brassica juncea* L. Czern.) and rapeseed (*Brassica napus* L.) to copper stress. Ecotoxicol Environ Saf 94:179–189
26. Feil SB, Pii Y, Valentinuzzi F, Tiziani R, Mimmo T, Cesco S (2020) Copper toxicity affects phosphorus uptake mechanisms at molecular and physiological levels in *Cucumis sativus* plants. Plant Physiol Biochem 157:138–147
27. Garcia JS, Dalmolin ÂC, Cortez PA, Barbeira PS, Mangabeira PA, França MG (2018) Short-term cadmium exposure induces gas exchanges, morphological and ultrastructural disturbances in mangrove *Avicennia schaueriana* young plants. Mar Pollut Bull 131:122–129
28. Gong Q, Wang L, Dai T, Zhou J, Kang Q, Chen H, Li K, Li Z (2019) Effects of copper on the growth, antioxidant enzymes and photosynthesis of spinach seedlings. Ecotoxicol Environ Saf 171:771–780
29. González-Mendoza D, GIL FEY, Escoboza-Garcia F, Santamaría JM, Zapata-Perez O (2013) Copper stress on photosynthesis of black mangle (*Avicennia germinans*). Anais da Academia Brasileira de Ciências 85(2):665–670
30. Grillo R, Mattos BD, Antunes DR, Forini MM, Monikh FA, Rojas OJ (2021) Foliage adhesion and interactions with particulate delivery systems for plant nanobionics and intelligent agriculture. Nano Today 37:101078
31. Gümüşsoy A, Başyiğit M, Kart EU (2023) Economic potential and environmental impact of metal recovery from copper slag flotation tailings. Resour Policy 80:103232
32. Ha N, Seo E, Kim S, Lee SJ (2021) Adsorption of nanoparticles suspended in a drop on a leaf surface of *Perilla frutescens* and their infiltration through stomatal pathway. Sci Rep 11(1):11556
33. Hong J, Peralta-Videa JR, Rico C, Sahi S, Viveros MN, Bartonjo J, Zhao L, Gardea-Torresdey JL (2014) Evidence of translocation and physiological impacts of foliar applied CeO_2 nanoparticles on cucumber (*Cucumis sativus*) plants. Environ Sci Technol 48(8):4376–4385
34. Hong YJ, Liao W, Yan ZF, Bai YC, Feng CL, Xu ZX, Xu DY (2020) Progress in the research of the toxicity effect mechanisms of heavy metals on freshwater organisms and their water quality criteria in China. J Chem 2020(1):9010348
35. Hossain MS, Abdelrahman M, Tran CD, Nguyen KH, Chu HD, Watanabe Y, Hasanuzzaman M, Mohsin SM, Fujita M, Tran LSP (2020) Insights into acetate-mediated copper homeostasis and antioxidant defense in lentil under excessive copper stress. Environ Pollut 258:113544
36. Ibrahim AS, Ali GA, Hassanein A, Attia AM, Marzouk ER (2022) Toxicity and uptake of CuO nanoparticles: evaluation of an emerging nanofertilizer on wheat (*Triticum aestivum* L.) plant. Sustainability 14(9):4914
37. Jin HJ, Wang HX (2019) Plant absorption and tolerance mechanism to heavy metal cadmium: research progress. Chin Agric Sci Bull 35(24):52–57
38. Joshi A, Sharma A, Bachheti RK, Husen A, Mishra VK (2019) Plant-mediated synthesis of copper oxide nanoparticles and their biological applications. In: Nanomaterials and plant potential. Springer, Cham, pp 221–237. https://doi.org/10.1007/978-3-030-05569-1_8
39. Jung Y, Ha M, Lee J, Ahn YG, Kwak JH, Ryu DH, Hwang GS (2015) Metabolite profiling of the response of burdock roots to copper stress. J Agric Food Chem 63(4):1309–1317
40. Kapoor D, Singh S, Kumar V, Romero R, Prasad R, Singh J (2019) Antioxidant enzymes regulation in plants in reference to reactive oxygen species (ROS) and reactive nitrogen species (RNS). Plant gene 19:100182

41. Kupper H, Götz B, Mijovilovich A, Kupper FC, Meyer-Klaucke W (2009) Complexation and toxicity of copper in higher plants. I. Characterization of copper accumulation, speciation, and toxicity in *Crassula helmsii* as a new copper accumulator. Plant Physiol 151(2):702–714
42. Lange B, van Der Ent A, Baker AJ, Echevarria G, Mahy G, Malaisse F, Meerts P, Pourret O, Verbruggen N, Faucon MP (2017) Copper and cobalt accumulation in plants: a critical assessment of the current state of knowledge. New Phytol 213(2):537–551
43. Li Y, Shi S, Zhang Y, Zhang A, Wang Z, Yang Y (2023) Copper stress-induced phytotoxicity associated with photosynthetic characteristics and lignin metabolism in wheat seedlings. Ecotoxicol Environ Saf 254:114739
44. Liao W, Zhu Z, Feng C, Yan Z, Hong Y, Liu D, Jin X (2023) Toxicity mechanisms and bioavailability of copper to fish based on an adverse outcome pathway analysis. J Environ Sci 127:495–507
45. Liu J, Dhungana B, Cobb GP (2018) Environmental behavior, potential phytotoxicity, and accumulation of copper oxide nanoparticles and arsenic in rice plants. Environ Toxicol Chem 37(1):11–20
46. Liu J, Shi X, Qian M, Zheng L, Lian C, Xia Y, Shen Z (2015) Copper-induced hydrogen peroxide upregulation of a metallothionein gene, OsMT2c, from *Oryza sativa* L. confers copper tolerance in Arabidopsis thaliana. J Hazard Mater 294:99–108
47. Liu J, Wen XS, Lang AD (2007) Progress in components and effects of plant root exudates. Food Drug 17:63–65
48. López-Moreno ML, de la Rosa G, Hernández-Viezcas JÁ, Castillo-Michel H, Botez CE, Peralta-Videa JR, Gardea-Torresdey JL (2010) Evidence of the differential biotransformation and genotoxicity of ZnO and CeO_2 nanoparticles on soybean (*Glycine max*) plants. Environ Sci Technol 44(19):7315–7320
49. Lv J, Christie P, Zhang S (2019) Uptake, translocation, and transformation of metal-based nanoparticles in plants: recent advances and methodological challenges. Environ Sci Nano 6(1):41–59
50. Lyubenova L, Kuhn AJ, Höltkemeier A, Schröder P (2013) Root exudation pattern of *Typha latifolia* L. plants after copper exposure. Plant Soil 370:187–195
51. Ma X, Yan J (2018) Plant uptake and accumulation of engineered metallic nanoparticles from lab to field conditions. Curr Opin Environ Sci Health 6:16–20
52. Ma X, Gurung A, Deng Y (2013) Phytotoxicity and uptake of nanoscale zero-valent iron (nZVI) by two plant species. Sci Total Environ 443:844–849
53. Marastoni L, Sandri M, Pii Y, Valentinuzzi F, Brunetto G, Cesco S, Mimmo T (2019) Synergism and antagonisms between nutrients induced by copper toxicity in grapevine rootstocks: monocropping vs. intercropping. Chemosphere 214:563–578
54. McDonald JP, Larson RR, Pahler LF, Thiese MS (2011) Real-time monitoring of arsenic, cadmium, copper and lead concentrations in workplace smelter particulates using a TSI 8520 DustTrak. Mining Metal Explor 28:13–20
55. Migocka M, Malas K (2018) Plant responses to copper: molecular and regulatory mechanisms of copper uptake, distribution and accumulation in plants. In: Plant micronutrient use efficiency. Academic Press, pp 71–86
56. Mir AR, Pichtel J, Hayat S (2021) Copper: uptake, toxicity and tolerance in plants and management of Cu-contaminated soil. Biometals 34(4):737–759
57. Morel JL, Mench M, Guckert A (1986) Measurement of Pb^{2+}, Cu^{2+} and Cd $^{2+}$ binding with mucilage exudates from maize (*Zea mays* L.) roots. Biol Fertil Soils 2:29–34
58. Mottaleb SA, Hassan AZ, El-Bahbohy R, Mahmoud AWM (2021) Are copper nanoparticles toxic to all plants? A case study on onion (*Allium cepa* L.). Agronomy 11(5):1006
59. Nie G, Zhao J, He R, Tang Y (2020) CuO nanoparticle exposure impairs the root tip cell walls of Arabidopsis thaliana seedlings. Water Air Soil Pollut 231:1–11
60. Noctor G, Mhamdi A, Chaouch S, Han YI, Neukermans J, Marquez-Garcia BELEN, Queval G, Foyer CH (2012) Glutathione in plants: an integrated overview. Plant Cell Environ 35(2):454–484

61. Panou-Filotheou H, Bosabalidis AM, Karataglis S (2001) Effects of copper toxicity on leaves of oregano (*Origanum vulgare* subsp. *hirtum*). Ann Botony 88(2):207–214
62. Peng C, Duan D, Xu C, Chen Y, Sun L, Zhang H et al (2015) Translocation and biotransformation of CuO nanoparticles in rice (*Oryza sativa* L.) plants. Environ Pollut 197:99–107
63. Pérez-de-Luque A (2017) Interaction of nanomaterials with plants: what do we need for real applications in agriculture? Front Environ Sci 5:12
64. Printz B, Lutts S, Hausman JF, Sergeant K (2016) Copper trafficking in plants and its implication on cell wall dynamics. Front Plant Sci 7:601
65. Quik JT, Lynch I, Van Hoecke K, Miermans CJ, De Schamphelaere KA, Janssen CR, Dawson KA, Stuart MA, Van De Meent D (2010) Effect of natural organic matter on cerium dioxide nanoparticles settling in model fresh water. Chemosphere 81(6):711–715
66. Rahman SU, Nawaz MF, Gul S, Yasin G, Hussain B, Li Y, Cheng H (2022) State-of-the-art OMICS strategies against toxic effects of heavy metals in plants: a review. Ecotoxicol Environ Saf 242:113952
67. Rahman SU, Wang X, Shahzad M, Bashir O, Li Y, Cheng H (2022) A review of the influence of nanoparticles on the physiological and biochemical attributes of plants with a focus on the absorption and translocation of toxic trace elements. Environ Pollut 310:119916
68. Rajput V, Minkina T, Ahmed B, Sushkova S, Singh R, Soldatov M, Laratte B, Fedorenko A, Mandzhieva S, Blicharska E, Musarrat J, Gorovtsov A (2020) Interaction of copper-based nanoparticles to soil, terrestrial, and aquatic systems: critical review of the state of the science and future perspectives. Rev Environ Contam Toxicol 252:51–96
69. Rajput V, Minkina T, Fedorenko A, Sushkova S, Mandzhieva S, Lysenko V, Duplii N, Fedorenko G, Dvadnenko K, Ghazaryan K (2018) Toxicity of copper oxide nanoparticles on spring barley (*Hordeum sativum* distichum). Sci Total Environ 645:1103–1113
70. Rajput V, Minkina T, Mazarji M, Shende S, Sushkova S, Mandzhieva S, Burachevskaya M, Chaplygin V, Singh A, Jatav H (2020) Accumulation of nanoparticles in the soil-plant systems and their effects on human health. Ann Agric Sci 65(2):137–143
71. Raliya R, Franke C, Chavalmane S, Nair R, Reed N, Biswas P (2016) Quantitative understanding of nanoparticle uptake in watermelon plants. Front Plant Sci 7:1288
72. Rashid MI, Shah GA, Sadiq M, Amin NU, Ali AM, Ondrasek G, Shahzad K (2023) Nanobiochar and copper oxide nanoparticles mixture synergistically increases soil nutrient availability and improves wheat production. Plants 12(6):1312
73. Rauser WE (1999) Structure and function of metal chelators produced by plants: the case for organic acids, amino acids, phytin, and metallothioneins. Cell Biochem Biophys 31:19–48
74. Saleem MH, Kamran M, Zhou Y, Parveen A, Rehman M, Ahmar S, Malik Z, Mustafa A, Anjum RM, Wang B, Liu L (2020) Appraising growth, oxidative stress and copper phytoextraction potential of flax (*Linum usitatissimum* L.) grown in soil differentially spiked with copper. J Environ Manag 257:109994
75. Shabbir Z, Sardar A, Shabbir A, Abbas G, Shamshad S, Khalid S, Murtaza G, Dumat C, Shahid M (2020) Copper uptake, essentiality, toxicity, detoxification and risk assessment in soil-plant environment. Chemosphere 259:127436
76. Shahid M, Pourrut B, Dumat C, Nadeem M, Aslam M, Pinelli E (2014) Heavy-metal-induced reactive oxygen species: phytotoxicity and physicochemical changes in plants. Rev Environ Contam Toxicol 232:1–44
77. Sheldon AR, Menzies NW (2005) The effect of copper toxicity on the growth and root morphology of Rhodes grass (*Chloris gayana* Knuth.) in resin buffered solution culture. Plant Soil 278:341–349
78. Siddiqi KS, Husen A (2020) Current status of plant metabolite-based fabrication of copper/copper oxide nanoparticles and their applications: a review. Biomater Res 24(11):1–15. https://doi.org/10.1186/s40824-020-00188-1
79. Su C, Chen A, Liang W, Xie W, Xu X, Zhan X, Zhan W, Peng C (2024) Copper-based nanomaterials: opportunities for sustainable agriculture. Sci Total Environ 171948
80. Sun L, Song F, Zhu X, Liu S, Liu F, Wang Y, Li X (2021) Nano-ZnO alleviates drought stress via modulating the plant water use and carbohydrate metabolism in maize. Arch Agron Soil Sci 67(2):245–259

81. Sunda WG (1975) The relationship between cupric ion activity and the toxicity of copper to phytoplankton. Doctoral dissertation, Massachusetts Institute of Technology
82. Tan JJ (2014) EDDS Alleviates the effects of excessive copper on the physiology biochemistry and epigenetics of rice. Wuhan University, Wuhan, China
83. Tripathi DK, Singh S, Singh S, Pandey R, Singh VP, Sharma NC, Prasad SM, Dubey NK, Chauhan DK (2017) An overview on manufactured nanoparticles in plants: uptake, translocation, accumulation and phytotoxicity. Plant Physiol Biochem 110:2–12
84. Tripathi GD, Javed Z, Dashora K (2024) Toxicity of copper oxide nanoparticles on agriculturally important soil rhizobacteria bacillus megaterium. Emerg Contam 10(1):100280
85. Wang ZC, Chen MX, Yang YX, Fang X, Liu ZJ, Wang LY, Ge MQ, Zhang C, Fang JG, Shangguan LF (2021) Effects of copper stress on plant growth and advances in the mechanisms of plant tolerance research. J Plant Nutr Fertilizers 27(10):1849–1863
86. Wassie AT, Ahmed IN, Bachheti A, Husen A, Bachheti RK (2024) Plant extract-assisted green synthesis and characterization of copper nanocomposite. In: Metal and metal-oxide based nanomaterials: synthesis, agricultural, biomedical and environmental interventions. Springer, Singapore, pp 43–66. https://doi.org/10.1007/978-981-99-7673-7_3
87. Xie C, Ma Y, Yang J, Zhang B, Luo W, Feng S, Zhang J, Wang G, He X, Zhang Z (2019) Effects of foliar applications of ceria nanoparticles and $CeCl_3$ on common bean (*Phaseolus vulgaris*). Environ Pollut 250:530–536
88. Xiong T, Zhang T, Xian Y, Kang Z, Zhang S, Dumat C, Shahid M, Li S (2021) Foliar uptake, biotransformation, and impact of CuO nanoparticles in *Lactuca sativa* L. var. ramosa Hort. Environ Geochem Health 43:423–439
89. Xu E, Liu Y, Gu D, Zhan X, Li J, Zhou K, Zhang P, Zou Y (2024) Molecular mechanisms of plant responses to copper: from deficiency to excess. Int J Mol Sci 25(13):6993
90. Xu Q, Qiu H, Chu W, Fu Y, Cai S, Min H, Sha S (2013) Copper ultrastructural localization, subcellular distribution, and phytotoxicity in *Hydrilla verticillata* (Lf) Royle. Environ Sci Pollut Res 20:8672–8679
91. Xuerong D, Fu Y, Huang Q, Xu Y, Zheng S, Sun Y (2023) Comparative effects of copper nanoparticles and copper oxide nanoparticles on physiological characteristics and mineral element accumulation in *Brassica chinensis* L. Plant Physiol Biochem 196:974–981
92. Yang Z, Chen J, Dou R, Gao X, Mao C, Wang L (2015) Assessment of the phytotoxicity of metal oxide nanoparticles on two crop plants, maize (*Zea mays* L.) and rice (*Oryza sativa* L.). Int J Environ Res Public Health 12(12):15100–15109
93. Yang Z, Xiao Y, Jiao T, Zhang Y, Chen J, Gao Y (2020) Effects of copper oxide nanoparticles on the growth of rice (*Oryza sativa* L.) seedlings and the relevant physiological responses. Int J Environ Res Public Health 17(4):1260
94. Zhang N, Xue C, Wang K, Fang Z (2020) Efficient oxidative degradation of fluconazole by a heterogeneous Fenton process with Cu-V bimetallic catalysts. Chem Eng J 380:122516
95. Zhang Z, Ke M, Qu Q, Peijnenburg WJGM, Lu T, Zhang Q, Qian H et al (2018) Impact of copper nanoparticles and ionic copper exposure on wheat (*Triticum aestivum* L.) root morphology and antioxidant response. Environ Pollut 239:689–697
96. Zheng J, Li Q, Zheng X (2023) Ocean acidification increases copper accumulation and exacerbates copper toxicity in *Amphioctopus fangsiao* (Mollusca: Cephalopoda): A potential threat to seafood safety. Sci Total Environ 891:164473

Role of Copper Nanomaterials for Pesticide Detection

Kundan Kumar Chaubey, Durgesh Wadhwa, Minerva Sharma, D. S. Tomar, Hridesh Kumar, Ashwani Kumar Sanghi, Vineet Kumar, Shailendra Thapliyal, and Ishita Yadav

Abstract The extensive application of insecticides in farming creates major environmental and health dangers, stressing the importance of sensitive, selective and cost-effective detection techniques. Due to their distinct physical and chemical features, copper nanomaterials (CuNMs) have emerged as a viable pesticide detection method. This chapter investigates the function of Copper nanoparticles (CuNPs) in pesticide detection, focusing on their advantages, detection processes, and applications. We cover CuNPs' applications in electrochemical detection with surface-enhanced Raman spectroscopy (SERS), fluorescence-based detection, and their potential to detect specific pesticides in varied matrices. In recent advancements CuNPs, CuNP-based colorimetric sensors have been developed for the detection of dithiocarbamate pesticides. CuNPs-based colorimetric indicators have been developed to detect pesticide residues in food and crops. CuNPs are also involved in the creation of affordable, approachable pesticide biosensing technologies. The challenges and future directions of CuNPs-based pesticide detection are discussed, focusing on scalability, toxicity, and integration with developing technologies. This chapter gives an extensive

K. K. Chaubey (✉) · M. Sharma · H. Kumar
Department of Biotechnology, School of Basic and Applied Sciences, Sanskriti University, Mathura, India
e-mail: kundan2006chaubey@gmail.com

D. Wadhwa
Department of Chemistry, School of Basic and Applied Sciences, Sanskriti University, Mathura, India

D. S. Tomar
Department of Physics, School of Basic and Applied Sciences, Sanskriti University, Mathura, India

A. K. Sanghi · V. Kumar
School of Allied Health Sciences, MVN University, Palwal, Haryana, India

S. Thapliyal
Uttaranchal Institute of Management, Uttaranchal University, Dehradun, India

I. Yadav
Institute of Agriculture Sciences, SAGE University Indore, Indore, India

R. K. Bachheti et al. (eds.), *Copper-Based Nanomaterial for Agricultural Practices*, Smart Nanomaterials Technology, https://doi.org/10.1007/978-981-96-8610-0_4

overview of CuNPs-based pesticide detection, emphasizing its potential to enhance food security, environmental monitoring and Sustainable agriculture.

Keywords Copper nanoparticles · Pesticide detection · Surface-enhanced Raman spectroscopy · Electrochemical detection · Fluorescence-based detection · Food safety

1 Introduction

Pesticides secure crops from bugs, infections, and weeds and raise efficiency per hectare. Since 1960, the production of key crops has more than tripled, mainly due to pesticides [33]. The use of pesticides is increasing day by day to ensure crop productivity and protect the crop from pests. It has also raised our concerns about the presence of their residues in polluting the environment, food, and water, which can affect human health, animals, and our ecosystem. In the field of nanotechnology, comprehending copper nanomaterials (CuNMs) has grown in importance and has significant ramifications for many other fields. The significance of tiny but potent particles is shown by research on the discovery of CuNMs. Copper nanoparticles (CuNPs) possess a unique combination of physical and chemical characteristics that facilitate their identification through various techniques, including spectroscopy, surface modification, electrochemical detection, and microscopy. These tests impact numerous fields, including data analytics, healthcare, environmental monitoring, and more. These initiatives have greatly benefited from the analysis of copper nanoparticles, which has ushered in a new era of nanotechnology advancement and provided us with the means to measure and regulate possible environmental issues [36, 45]. In addition to electronics and energy storage, it supports several businesses and catalysis. By guaranteeing that the items are safe and adhere to the law, it has also won the trust of consumers. In the field of nanotoxicology, the identification of copper nanomaterials promotes the quest for safety and health.

Chromatographic and spectroscopic methods are the traditional pesticide detection methods, but they are time-consuming, expensive, and require expertise. So, it's time to develop the nanomaterial-based sensors that can rapidly detect the pesticide. Out of other nanomaterials, copper-based nanomaterials are more eligible for use in pesticide detection. CuNMs have unique optical, electronic, and catalytic properties. Nano-biotechnology is a field where we can study the potential of different nanomaterials in the protection of plants from pests [3]. Copper oxide nanoparticles have been synthesized on copper wires to detect hydrogen peroxide, showcasing high sensitivity and selectivity. *Phragmanthera austroarabica* is extracted as part of the green synthesis process to create Copper nanoparticles (CuNPs) [1]. The advancement in nanomaterials (NMs) production highlights the promising applications of copper nanomaterials in pesticide detection for sustainable agro-ecosystems [3]. CuNMs have emerged as a vital component in various industrial applications including electronics, catalysis, and renewable energy technologies [13]. Their unique properties,

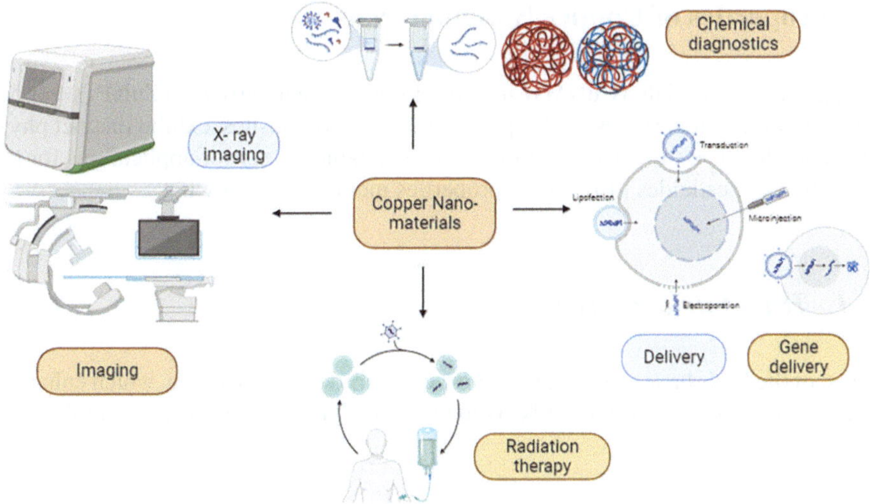

Fig. 1 CuNMs potential applications

such as high electrical conductivity, excellent thermal conductivity, and catalytic activity, make NMs essential in developing pesticide detection. Before using NMs in various applications, the effective detection and analysis of copper nanomaterial are crucial for ensuring their quality, functionality, and safety in numerous applications, as shown in Fig. 1.

Copper nanomaterials, such as copper oxide nanorods (CuO NRs) and copper oxide nanoparticles (CuO), have been utilized in the fabrication of electrodes for pesticide detection [30, 35]. CuO NR sensors showed high sensitivity and selectivity in identifying different organophosphate pesticides (OPPs) and separating them from interference molecules [35]. Additionally copper nanoparticles capped with Tween 80 demonstrated a cost-effective colorimetric probe for selective detection of thiram pesticide with low detection limits [2]. Copper nanomaterial-based sensors offer promising platforms for the sensitive and reliable detection of pesticide residues in various environmental samples. Traditional pesticide detection methods, such as gas chromatography (GC) and high-performance liquid chromatography (HPLC) coupled with mass spectrometry (MS), face several limitations, including complex sample preparation, high equipment costs, the need for skilled personnel, lengthy analysis times, and unsuitability for on-site rapid detection. These drawbacks make them impractical for extensive field applications and the rapid screening of pesticide residues in food and other agricultural products [17]. The aim of this chapter is to provide an overview of copper nanomaterials for pesticide detection

2 Properties of Copper Nanomaterials

Copper nanomaterials (CuNMs) including copper nanoparticles (CuNPs), copper oxide (CuO) nanoparticles, and copper-based nanocomposites, exhibit distinct physical and chemical properties that make them suitable for sensor applications for the detection of pesticides [40]. These properties include the following:

2.1 High Surface Area

Most of the NMs have a large surface area compared to their volume, which enhances the reactivity and sensitivity of NMs to target analytes such as pesticide molecules.

2.2 Catalytic Activity

CuNMs have significant catalytic properties, which support them in sensing applications. For example, CuO nanoparticles can catalyze specific reactions, which start altering the system's properties and characteristics in response to pesticides and other materials.

2.3 Optical Properties

CuNMs, particularly those with a plasmonic response, can change their optical properties, such as absorption and scattering, upon interaction with pesticides and other materials. These properties make CuNMs ideal for optical sensing techniques during pesticide detection.

2.4 Electrochemical Properties

Due to CuNMs' electrochemical properties, they are widely used in the development of electrochemical sensors. They can enhance electron transfer in reactions, which can improve the sensitivity and selectivity of detection systems used for pesticides and other chemical compounds.

These properties of CuNMs make them highly suitable for the development of sensors that can easily detect pesticides and other materials with high sensitivity and selectivity. They also make them very beneficial in a wide range of applications [4, 5].

3 Mechanisms of Pesticide Detection Using Copper Nanomaterial

Copper nanomaterials-based sensors can be employed in various detection methods for pesticides and other chemical compounds using colorimetric, electrochemical, and optical sensing techniques. Below are the key mechanisms used with CuNMs to detect pesticides (Fig. 2).

3.1 Colorimetric Detection

Copper nanomaterials synthesized with glucose as a reducing agent function as a fluorescent probe for the colorimetric detection of Ag^+ ions in water [42]. Silver ions suppress the Glc-CuNPs' fluorescence, making visible color changes for detection. The nanomaterial sensor exhibits good selectivity for Ag^+ ions and can detect concentrations in the range of 100–600 mol/L with a linear relationship. This method offers a simple and effective way to detect heavy metal silver ions, with promising applications in environmental detection. When CuNPs are exposed to pesticides and other materials, they can undergo changes in their size, shape, or aggregation, which leads to a shift/change in their optical properties and characteristics, particularly in the UV-visible spectrum. For example, aggregation of CuNPs may occur due to pesticide molecules, which results in the color change being identified and quantified by spectrophotometer [63]. These color changes are often found due to the interaction of

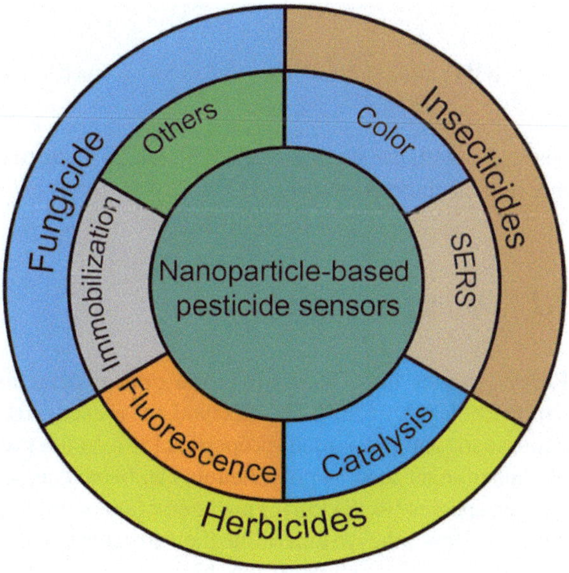

Fig. 2 The functions of NPs and their application in pesticide detection [56]

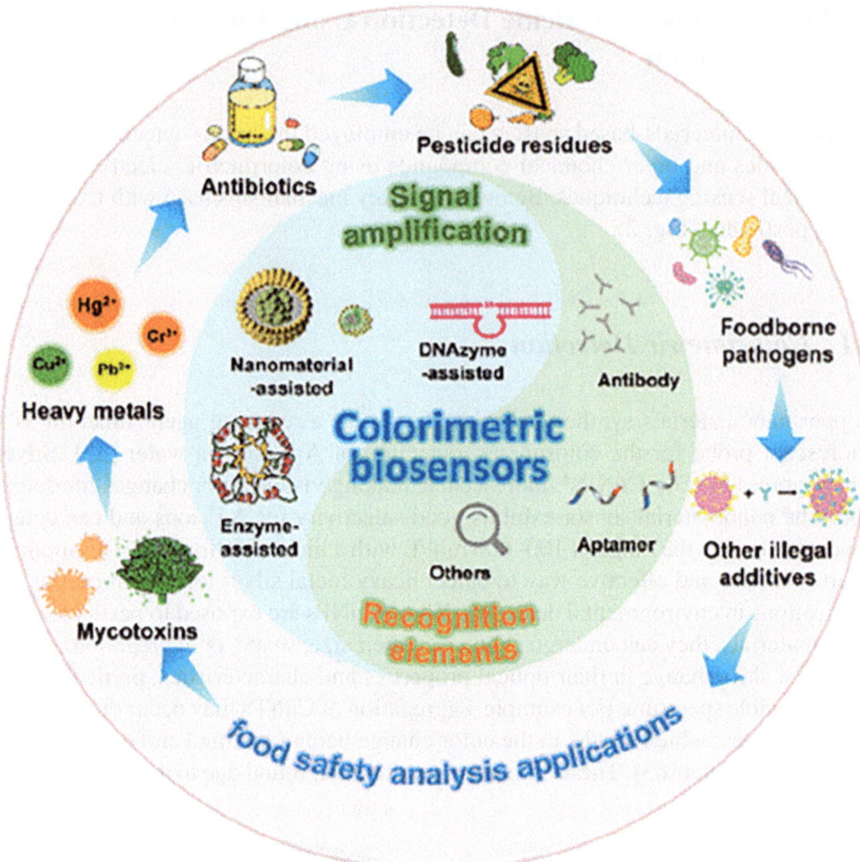

Fig. 3 NMs based colorimetric biosensors for agri-food safety analysis [62]

pesticides with the surface of CuNPs, which leads to a measurable signal developed due to their interaction (Fig. 3).

3.2 *Optical Detection*

Recent studies have focused on the optical detection of pesticides using copper nanoparticles [6, 24, 25]. The mechanism is based on the interaction of nanomaterials' antioxidant and prooxidant properties [38]. In the presence of copper ions, antioxidant chemicals such as polyphenols can behave as prooxidants, increasing reactive oxygen species (ROS) [38]. This interaction is investigated using electron spin resonance (ESR) studies to determine their structures and ability to cause oxidative stress. The study also investigates how high quantities of antioxidants and specific metal ions, such as copper, can produce considerable amounts of ROS in fake intestinal juice [38]. This underscores the significance of knowing the physiological consequences of these interactions: Fluorescence, absorbance, and surface-enhanced Raman spectroscopy (SERS). The SERS has shown that it can provide a very strong signal while providing the same information as standard Raman Spectroscopy. Raman signals can be improved using SERS spectroscopy.

3.3 *Electrochemical Detection*

An effective and noninvasive technique called electrochemical impedance spectroscopy (EIS) can analyze external electrical properties. Any building material or specific process that may affect the performance, protection or capability of the electrochemical process can be examined using EIS. According to Grieshaber et al. [7] EIS is a useful method for studying polymer degradation and designing and analyzing biosensor data transmission. By disrupting the polymer layer, the electric field can be tested directly or indirectly by coating it with an insulating polymer and subjecting it to enzymatic treatment. If this process causes free ions to enter the polymer, the insulation quality of the polymer will be affected, thus changing the impedance characteristics of the switch. Electrochemical EIS has recently been used to calculate the design of hydrogen peroxide (HRP) biosensors. The impedance method can be used in electrochemical sensing to observe changes in electrical properties from biometric events that alter the electrode surface. Copper-based electrodes are used in the development of electrochemical sensors to detect pesticides. The presence and availability of pesticides may interfere. They can cause a change in the electrochemical responses like changes in current or potential of copper electrodes, which can be used to know pesticide concentrations. CuO NMs have strong catalytic properties that help to improve the pesticide residue detection capacity in complex matrices [52].

3.4 *Fluorescence-Based Detection*

Some CuNMs, especially copper nanoclusters, contain fluorescence properties that can detect any chemical compounds very easily. When specific pesticides interact with CuNMs, the fluorescence intensity of these NMs can be modulated/ modified; these changes in the fluorescence intensity facilitate the development of a sensitive and rapid detection method. The interaction of CuNMs and pesticides is often characterized using changes in wavelength and fluorescence intensity using different fluorometric techniques [27].

3.5 *Surface-Enhanced Raman Spectroscopy (SERS)*

CuNMs contain plasmonic properties, which are also useful in SERS for detecting of pesticides detecting pesticides and other chemicals. The high local electromagnetic field created near the CuNMs surface can enhance Raman signals, making this technique suitable and helpful for detecting low pesticide concentrations [14].

3.6 *Bio-sensing and Immunoassays*

The majority of biological samples that are removed in order to test for certain biomarkers show up as extremely complicated media with a wide range of variable concentrations of the various bio-molecules that make them up [19]. NPs link the bulk material, atoms, and tiny molecules. A number of variables, including the angle of curvature, these nanoscale objects exhibit distinct atomic organization and behaviors on their surface, resulting in various nanoscale phenomena. Because of this, they exhibit unique optical, mechanical, and magnetic characteristics that are highly reliant on their shape and size and are missing in the bulk form. Constrained surface plasmon resonance bands, enhanced fluorescence, restricted emission peaks, increased scattering properties (gold and silver NPs), high photo-stability (quantum dots), catalytic activities (Graphene Oxide), and superparamagnetism are some of these traits.

3.7 *Environmental Monitoring*

Environment monitoring (EM) includes managing and planning disasters, reducing pollutants, and successfully resolving issues that occur from unhealthful environmental circumstances. Environmental medicine addresses air and water pollution, dangerous radiation, shifting weather patterns, earthquakes, and more. Numerous variables, some artificial and others resulting from natural causes, contribute to the sources of pollution. EM's mission is to accurately handle these difficulties to maintain the healthy environment for a planet and civilization [39]. Among the many uses are weather forecasting, agricultural damage assessment, water quality control, and air pollution control, which environmental monitoring techniques are used. The goal is to create an environment that is conducive to agriculture, human habitation, or any combination of these.

3.8 Quantum Dots and Nanosensors

A sensor is an apparatus that can change a system's chemical, biological, or physical components into a signal that is valuable, processable, and measured analytically using a transducer [16]. If a sensor interacts at the nanoscale, it is called a nanosensor. Researchers have focused much interest on quantum dots due to their distinct thermal, electrical, optical, and catalytic qualities. These properties have led to the development of numerous electrochemical sensors.

4 Applications of CuNMs in Pesticide Detection

Due to the unique properties of CuNMs, such as high surface area, excellent electrical conductivity, and catalytic activity, CuNMs have shown substantial potential in pesticide detection. CuNMs' ability to interact with pesticides and other chemicals and undergo chemical changes when interacting with specific pesticides makes them unique and effective for developing sensitive, specific, and trustworthy sensors for pesticide detection. The CuNMs have some unique properties which have been exploited in various applications for pesticide detection, including (Table 1).

4.1 Agriculture Field

Pesticides are commonly employed in agriculture, but their long-term presence in the environment endangers humans, animals, insects, and ecosystems. Pesticides used excessively for pest control, crop yield enhancement, and other purposes leave a considerable environmental residue. Pesticide monitoring and control technologies have been created employing a variety of strong, dependable, and reusable composites. Researchers revealed that the many characteristics of carbon nanomaterials, such as their high porosity, conductivity, and ease of electron transfer, make them suitable for effectively detecting pesticide residues in food. The role of carbon nanomaterials in the analysis of pesticide residues in different food matrices. Carbon nanotubes and carbon dots are examples of carbon nanomaterials [31].

4.2 Insecticides

Insecticides are types of pesticides developed to harm, repel, kill, or mitigate one or more species of insects to increase agricultural productivity around the globe. Many of the insecticides are toxic and harmful to humans, animals, and the environment. Here, we will discuss CuNPs based sensors for detecting pesticides like

Table 1 Types of CuNMs, their roles, mechanism of action and advantages

Type of CuNMs	Role in pesticide detection	Mechanism of action	Advantages	References
CuNPs	Pesticide residues in food and water	CuNPs based sensor interact with pesticide, leads to change in optical properties like color or fluorescence	High surface area, sensitivity, easy synthesis	Zheng et al. [59]
CuO NPs	Organophosphate and carbamate pesticides	CuO interacts with pesticides, causing a measurable change in conductivity or electrochemical response	High stability, selective interaction with organophosphates	Yu et al. [51]
CuS NPs	Organophosphates and other toxic chemicals	CuS nanomaterials enhance fluorescence and photoluminescence upon interaction with pesticide molecules	High sensitivity, simple detection methods	Wang et al. [44]
CuNWs	Pesticides in environmental samples	CuNWs enable rapid electrochemical detection due to their excellent electrical conductivity and large surface area	Fast response, high reproducibility, and cost-effectiveness	Li et al. [23]
CuNPs/ Graphene	wide range of pesticides including neonicotinoids	Copper nanoparticle-graphene composites enhance sensitivity and conductivity, allowing for faster and more sensitive detection	High stability, enhanced conductivity, and greater sensor performance	Zhang et al. [55]
Copper-based thin films	Pesticides in agricultural products	Thin copper films interact with pesticide molecules, leading to a change in surface plasmon resonance (SPR) or other optical signals	High precision and selective sensing capability	Kumar et al. [18]

organophosphorus, new nicotine, pyrethroids, and organochlorine that are widely used in current agriculture. By pesticide testing, we can eliminate substandard and low-category products contaminated with pesticides and other harmful chemicals and ensure food safety [11]. So, monitoring of pesticide residues in food is critical [41] (Table 2).

As we know, CuNP-based sensors can detect organophosphate pesticides (e.g., paraoxon), which prevent acetylcholinesterase (AChE) activity. CuNP-based sensors

Table 2 Copper nanomaterials for the detection of insecticides: a detailed table

Method	Nanomaterial type	Insecticide detected	Detection limit	Advantages	Performance	References
Electrochemical sensing	Copper nanoparticles (CuNPs)	Organophosphates (e.g., parathion, chlorpyrifos)	0.5 nM	High sensitivity, rapid response, low cost, ease of fabrication	High sensitivity and selectivity for organophosphates with a wide linear range. Good stability	Zhang et al. [28], Liu et al. [54]
Colorimetric sensing	Copper nanoclusters (CuNCs)	Pesticides (e.g., carbamate, organophosphates)	10 nM	Simple, inexpensive, no need for sophisticated equipment, fast response	Shows a clear color change in the presence of insecticides, suitable for on-site testing	Wang et al. [43]
Surface-enhanced raman spectroscopy (SERS)	Copper nanoparticles (CuNPs)	Pyrethroids (e.g., permethrin)	1 μM	High sensitivity, non-destructive, minimal sample preparation	Excellent sensitivity for pyrethroid detection, rapid analysis, suitable for field applications	Yang et al. [50]
Fluorescence sensing	Copper nanoclusters (CuNCs)	Chlorpyrifos, imidacloprid	0.1 μM	High sensitivity, easy to use, real-time monitoring, can be miniaturized	CuNC-based sensors show strong fluorescence quenching in the presence of insecticides, allowing for detection at low concentrations	Xie et al. [48]
Electrochemical sensor array	Copper oxide nanostructures (CuO)	Acaricides, organophosphates	0.1 μM	Multi-analyte detection, high selectivity, can detect multiple insecticides simultaneously	Excellent performance in real-world conditions with high reproducibility. Can detect multiple insecticides in a single assay	He et al. [10]
Electrochemical impedance spectroscopy	Copper nanostructures (CuNWs)	Chlorpyrifos, dimethoate	2.5 nM	High sensitivity, fast response, robust for complex matrices	High selectivity for specific insecticides, with low interference from other compounds	Zhang et al. [57]
Photoelectrochemical sensing	Copper sulfide nanoparticles (CuS)	Organophosphates, carbamates	0.01 nM	High sensitivity, low detection limit, suitable for real-time detection	High performance under low-light conditions, suitable for use in environmental monitoring and field applications	Li et al. [22]

(continued)

Table 2 (continued)

Method	Nanomaterial type	Insecticide detected	Detection limit	Advantages	Performance	References
Surface plasmon resonance (SPR)	Copper nanostructures (CuNPs)	Malathion, diazinon	0.2 μM	High specificity, real-time monitoring, label-free detection	High precision in detecting organophosphate insecticides in water and soil samples	Zhou et al. [61]
Optical sensing	Copper nanodisks (CuNDs)	Carbamate, organophosphates	2 nM	High sensitivity, cost-effective, and simple fabrication technique	Excellent stability and reproducibility, can be used for rapid detection in various environmental samples	Liu et al. [29]
Magnetic sensor-based detection	Copper nanoparticles (CuNPs)	Pyrethroids, organophosphates	0.5 nM	High selectivity, easy to operate, cost-effective, can be integrated into portable devices	High affinity for target insecticides with quick and reliable signal generation. Suitable for on-site detection	Jang et al. [12]

can also monitor this inhibition directly or indirectly through electrochemical responses [60].

4.3 Detection of Organophosphates

Organophosphates are prevalent pesticides among the most widely used pesticides nowadays [37]. Organophosphorus is an ester of phosphoric acid that contains oxygen, nitrogen, carbon, and sulfur compounds [8]. Several published studies have focused on CuNMs for detecting residues of organophosphate pesticides in food, water, and soil. CuNMs and CuO NPs have been functionalized using different polymers and biomolecules to improve their identification capacity and selectivity for detecting organophosphate and their related compounds [52]. Wannasri et al. [47] have used a facile nonenzymatic electrochemical sensor developed using CuO NPs deposited on activated carbon for the highly sensitive detection of methyl parathion.

4.4 Detection of Herbicides and Fungicides

Glyphosate and Chlorothalonil are very common Herbicides and fungicides, respectively. There is a long range of pesticides detected using copper-based sensors. CuNMs have been used to develop these types of sensors, which have high sensitivity and specificity and can detect very low concentrations of these types of chemicals and pesticides we use in agricultural runoff and environmental samples [21]. Some NPs, such as MWCNTs and copper nanowires, having electrocatalytic abilities, can greatly escalate the sensitivity of electrochemical detection of pesticides like trifluralin [9, 32].

4.5 On-Site Detection

CuNMs based sensors are very robust and versatile and they have been developed for rapid, on-site, field-based detection of pesticides. These sensors are portable, having low cost, and ease of use of these sensors make them very ideal for on-the-spot detection and monitoring of pesticide contamination in agricultural settings [46]. Zeng et al. [58] developed a molecularly imprinted polymer containing copper-based metal-organic skeleton (MIP/HKUST-1) for the selective recognition of tetrabromobisphenol A (TBBPA). The adsorbed TBBPA was degraded due to the presence of hydrogen peroxide in the reaction. The hydrogen peroxide and HKUST-1 combined action reduced the color of oxTMB. This colorimetric strategy was ultrasensitive and highly selective for the detection of TBBPA.

4.6 Enzyme-Based Biosensors

CuNMs are frequently used in enzyme-based biosensors that act as catalysts to boost the reaction between the pesticide and the immobilized enzyme used for detection. These biosensors are highly sensitive and specific and can also detect low pesticide concentrations. Sensor's surface generally immobilized by Enzymes like acetylcholinesterase (AChE) or butyrylcholinesterase (BChE) to interact with organophosphates and undergo for the inhibition. Here, CuNMs facilitate strengthening the signal produced by this interaction of enzymes and pesticides, making the detection more promising and sensitive.

CuNMs facilitate the enzyme's improvement of electrocatalytic activity, which leads to more perceptible changes in the presence of pesticides.

4.7 Copper Nanowires and Nanotubes for Pesticide Detection

Copper nanowires (CuNWs) and nanotubes (CuNTs) have also been applied in pesticide sensors. Due to their elongated structures, CuNWs and CuNTs offer a larger surface area compared to nanoparticles, which allows for more binding sites for the pesticide molecules. This results in improved sensitivity and selectivity in detection [26]. Jiang et al. [15] have reported that CuNW-based sensors detected chlorpyrifos pesticides with high sensitivity and specificity and detected them at low concentrations.

4.8 Copper-Graphene Hybrid Nanomaterials

In copper-graphene hybrids, graphene gives excellent conductivity, and CuNMs give catalytic properties, offering additive synergistic effects that increase the performance, accuracy, and reliance of sensors in detecting pesticides. These materials contain high surface area and conductivity, making them ideal for detecting a wide range of pesticides. Zhang et al. [53] have developed a hybrid material of CuNPs and graphene oxide (GO) to detect carbamate pesticides. This hybrid material is very sensitive, specific, and reliable in detecting pesticides, achieving detection of low levels of concentration and a wide linear range.

4.9 Copper-Based Optical Sensors

CuNMs can also be used as optical sensors to detect pesticides through light absorption or fluorescence changes. The interaction between pesticides and CuNMs causes a change in optical properties like color or light scattering. These optical properties facilitate the measurement of the pesticide's presence and concentration. Copper oxide (CuO) nanostructures have unique optical properties that facilitate the colorimetric sensors that can be used to detect pesticides like methyl parathion [20].

5 Challenges and Future Directions

Several challenges remain here despite the promising potential of CuNMs in pesticide detection. These challenges remain:

5.1 Stability

CuNMs are very prone to oxidation; the presence of oxygen or moisture can alter the properties and characteristics of CuNMs and affect the reliability of CuNMs-based sensors. Various researchers are searching for and exploring strategies to stabilize the CuNMs for further alteration, such as fabricating/coating them with protective layers of different chemicals or using copper alloys [34].

5.2 Selectivity

As we know, CuNMs are sensitive to pesticides; to achieve high selectivity for a broad range of complex environmental samples like soil, water, and food, developing sensors became more challenging. Current research uses specific receptors or ligands during the development and functionalization of CuNMs to improve selectivity. However, it remains challenging, and more and more research is required to reach a new milestone in this area.

5.3 Integration into Commercial Systems

Most CuNMS-based sensors show promising results in laboratory conditions, but translating these sensors/technologies into our practical field-based study and commercially viable systems for extensive and widespread use remains a significant hurdle [49]. Issues like consistency, scalability, reproducibility, and various regulatory approvals must be discussed and addressed before these sensors can be widely adopted for further use in pesticide detection.

We can design a plan for further future research that could focus on overcoming these challenges by developing novel CuNMs-based sensor platforms that are more sensitive, specific, stable, selective, and non-reactive and integrated with advanced portable detection technologies, such as smartphone-based applications, for real-time monitoring of pesticide contamination in a wide range of samples.

6 Conclusion

CuNMs have promising characteristics that differentiate them from other NMs in developing sensors. CuNMs hold great potential for pesticide detection due to their unique properties like selectivity, catalytic activity, optical characteristics, and high surface area. CuNMs have already been successfully employed in various detection methods, such as electrochemical, colorimetric, fluorescence, and SERS-based techniques. Despite all available challenges for stability and selectivity, CuNMs-based sensors are the only option poised to play a significant and promising role in future of pesticide detection, offering sensitive, specific, rapid, on-site, and cost-effective solutions for environmental contamination and food safety-related monitoring.

Competing Interests The authors declare that they have no competing interest.

Declaration of Generative AI and AI-Assisted Technologies in the Writing Process During the preparation of this work the book chapter author(s) have not taken any assistance from AI, and the book chapter author(s) reviewed and edited the content as needed and take(s) full responsibility for the content of the publication. All authors read and approved the final version of the manuscript.

References

1. Akter M, Sikder MT, Rahman MM, Amin MO (2019) Recent trends in the synthesis, properties, and potential applications of copper nanoparticles. Mat Sci Eng 242:85–98
2. Anh NT, Dinh NX, Van Tuan H, Thuan TH, Tung LM, Le VP, Tri DQ (2021) Le AT (2021) Cost-effective Tween 80-capped copper nanoparticles for ultrasensitive colorimetric detection of Thiram pesticide in environmental water samples. J Nanomater 1:5513401
3. Chongsuebsirikul J (2019) Copper oxide nanorods electrodes for organophosphate pesticide sensor. In: 2019 IEEE 14th International conference on nano/micro engineered and molecular systems (NEMS), Bangkok, Thailand, pp 520–523
4. Gao Y, Zhuang L, Zhang Y, Zhang Y, Pei L, Li X (2024) Electrochemical performance of polyaniline-polyvanadate-copper nanomaterials for the detection of benzoic acid. Macromol Chem Phys 225(14):2300444
5. Gong Z, Huang Y, Hu X, Zhang J, Chen Q, Chen H (2023) Recent progress in electrochemical nano-biosensors for detection of pesticides and mycotoxins in foods. Biosensors 13(1):140
6. Gopal G, Roy N, Mukherjee A (2023) Recent developments in the applications of GO/rGO-based biosensing platforms for pesticide detection. Biosensors 13(4):488
7. Grieshaber D, MacKenzie R, Voros J, Reimhult E (2008) Electrochemical biosensors-sensor principles and architectures. Sensor 8:1400–1458

8. Gui WJ, Wang ST, Guo YR, Zhu GN (2008) Development of a one-step strip for the detection of triazophos residues in environmental samples. Anal Biochem 377(2):202–208
9. Haghighi M, Irandoust M, Shariati-rad M (2019) Simultaneous determination of antinonin and trifluralin by electrochemical method and net analyte signal interferent modeling. Microchem J 46:34–40
10. He X et al (2021) Electrochemical sensor array based on copper oxide nanostructures for multi-analyte detection of insecticides. Sens Actuators B: Chem 330:129391
11. Hua Z, Yu T, Liu D, Xianyu Y (2021) Recent advances in gold nanoparticles-based biosensors for food safety detection. Biosens Bioelectron 179:113076
12. Jang W et al (2022) Magnetic sensor-based detection of pyrethroids and organophosphates using copper nanoparticles. J Agric Food Chem 70(11):3407–3414
13. Joshi A, Sharma A, Bachheti RK, Husen A, Mishra VK (2019) Plant-mediated synthesis of copper oxide nanoparticles and their biological applications. In: Nanomaterials and plant potential. Springer, Cham, pp 221–237. https://doi.org/10.1007/978-3-030-05569-1_8
14. Jin Y, Xie J, Chen W (2020) Copper nanomaterials in surface-enhanced raman spectroscopy (SERS) for environmental monitoring. Anal Chem 92(7):4696–4704. https://doi.org/10.1021/acs.analchem.0c00299
15. Jiang et al (2018) Copper nanowire-based electrochemical sensors for pesticide detection. Sensors 18(7):2277
16. Karadurmus L, Ozcelikay G, Vural S, Ozkan SA (2021) An overview on quantum dot-based nanocomposites for electrochemical sensing on pharmaceutical assay. Iran J Pharm Res 20:187
17. Kumar A, Bakshi M (2021) Copper-based nanoparticles in the soil-plant environment: assessing their applications, interactions, fate and toxicity. Chemosphere 281:0045–6535
18. Kumar et al. (2020) Copper Thin Films for Real-Time Pesticide Detection via Surface Plasmon Resonance. Biosens Bioelectron 150:111878
19. Lara S, Perez-Potti A (2018) Applications of nanomaterials for immunosensing. Biosensors 8:104
20. Lee J et al (2020) Copper oxide nanostructures for optical detection of methyl parathion pesticide. Environ Sci Technol 54(22):14102–14108
21. Li X, Zhang L, Zhao D (2020) Electrochemical sensors based on copper nanomaterials for detection of herbicides. Sens Actuators B: Chem 303:127116. https://doi.org/10.1016/j.snb.2019.127116
22. Li S et al (2020) Photoelectrochemical sensing of organophosphate pesticides using copper sulfide nanoparticles. J Electroanal Chem 870:113688
23. Li et al (2021) Copper nanowires-based electrochemical sensor for pesticide detection. Electrochim Acta 370:137684
24. Li H, Murugesan A, Shoaib M, Sheng W, Chen Q (2024) Functionalized metal-organic frameworks with biomolecules for sensing and detection applications of food contaminants. Crit Rev Food Sci Nutr 18:1–33
25. Liang M, Guo LH (2009) Application of nanomaterials in environmental analysis and monitoring. J Nanosci Nanotechnol 9(4):2283–2289
26. Liu J et al (2016) A highly sensitive copper nanoparticle-based electrochemical sensor for organophosphate pesticide detection. J Hazard Mater 313:22–28
27. Liu J, Zhang X, Zhao X (2019) Fluorescent copper nanoclusters for detection of organophosphates. J Hazard Mater 375:1–8. https://doi.org/10.1016/j.jhazmat.2019.04.015
28. Liu Y et al (2021) Copper nanoparticles for the electrochemical detection of insecticides in agricultural and environmental samples. J Agric Food Chem 69(31):9157–9165
29. Liu W et al (2022) Optical detection of carbamate insecticides using copper nanodisks. Sensors 22(16):6148
30. Mahmoudpour M, Karimzadeh Z, Ebrahimi G, Hasanzadeh M, Ezzati Nazhad Dolatabadi J (2021) Synergizing functional nanomaterials with aptamers based on electrochemical strategies for pesticide detection: current status and perspectives. Crit Rev Anal Chem 52(8):1818–1845
31. Mishra S, Mishra S, Patel SS, Singh SP, Kumar P, Khan MA, Awasthi H, Singh S (2022) Carbon nanomaterials for the detection of pesticide residues in food: a review. Environ Pollut 310:119804

32. Mirabi-Semnakolaii A, Daneshgar P, Moosavi-Movahedi AA, Rezayat M, Norouzi P, Nemati A, Farhadi M (2011) Sensitive determination of herbicide trifluralin on the surface of copper nanowire electrochemical sensor. J Solid State Electrochem 15:1953–1961
33. Mohan CR, Kandasamy R, Kabiriyel J (2024) Correlation between electrical conductivity and antibacterial activity of chitosan-stabilized copper and silver nanoparticles. Carbohydr Polym Technol Appl 7:100503
34. Saha P, Pal S, Singh P (2020) Stabilization of copper nanoparticles for pesticide detection. J Nanopart Res 22(4):85. https://doi.org/10.1007/s11051-020-04845-x
35. Sakdarat P, Chongsuebsirikul J, Thanachayanont C, Prichanont S, Pungetmongkol P (2021) Development of a nonenzymatic electrochemical sensor for organophosphate pesticide detection using copper (II) oxide nanorod electrodes. J Nanomater 2021(1):6623668
36. Siddiqi KS, Husen A (2020) Current status of plant metabolite-based fabrication of copper/copper oxide nanoparticles and their applications: a review. Biomater Res 24(11):1–15. https://doi.org/10.1186/s40824-020-00188-1
37. Tan X, Yu C, Tang J, Wu W, Yang Q, Hou X (2022) Progress in nanomaterials-based enzyme and aptamer biosensor for the detection of organophosphorus pesticides. Crit Rev Anal Chem 54:247–268
38. Tasaki-Handa A (2021) Mechanisms of antioxidant and prooxidant interactions of nanomaterials in the presence of copper ions: implications for oxidative stress. J Nanosci Nanotechnol 21(4):2345–2356
39. Ullo SL, Sinha GR (2020) Advances in smart environment monitoring systems using IoT and sensors. Sensors 20:3113
40. Vrabelj T, Finsgar M (2022) Recent progress in non-enzymatic electroanalytical detection of pesticides based on the use of functional nanomaterials as electrode modifiers. Biosensors 12(5):263
41. Wahab S, Muzammil K, Nasir N, Khan MS, Ahmad MF, Khalid M, Ahmad W, Dawria A, Reddy LK, Busayli AM (2022) Advancement and new trends in analysis of pesticide residues in food: a comprehensive review. Plants 11(9):1106
42. Wang N, Ga L, Jia M, Ai J (2019) Synthesis of fluorescent copper nanoparticles and ultrasensitive free label detection of Ag+. J Nanomater 2019(1):4089731
43. Wang Q et al (2019) Colorimetric sensing of organophosphate pesticides based on copper nanoclusters. Talanta 199:286–292
44. Wang Z, Li S, Yang L (2021) On-site detection of pesticides using copper nanomaterials. Environ Chem Lett 19:2089–2097
45. Wassie AT, Ahmed IN, Bachheti A, Husen A, Bachheti RK (2024) Plant extract-assisted green synthesis and characterization of copper nanocomposite. In: Metal and metal-oxide based nanomaterials: synthesis, agricultural, biomedical and environmental interventions. Springer, Singapore, pp 43–66. https://doi.org/10.1007/978-981-99-7673-7_3
46. Wang X et al (2021) Copper sulfide nanomaterials for fluorescent sensing of pesticides. Sens Actuators B: Chem 335:129494
47. Wannasri N, Uppachai P, Butwong N, Jantrasee S, Isa IM, Loiha S, Srijaranai S, Mukdasai S (2022) A facile nonenzymatic electrochemical sensor based on copper oxide nanoparticles deposited on activated carbon for the highly sensitive detection of methyl parathion. J Appl Electrochem 52(3):595–606
48. Xie Z et al (2022) Fluorescent copper nanoclusters for selective detection of organophosphate pesticides. Anal Chim Acta 1194:58–66
49. Yadav A, Yadav K, Ahmad R, Abd-Elsalam KA (2023) Emerging frontiers in nanotechnology for precision agriculture: advancements, hurdles and prospects. Agrochemicals 2(2):220–256
50. Yang Y et al (2021) Surface-enhanced Raman scattering using copper nanoparticles for the detection of pyrethroid insecticides. J Raman Spectrosc 52(9):1687–1693
51. Yu et al (2019) Copper oxide nanoparticles for pesticide detection via electrochemical sensing. J Hazard Mater 369:35–43
52. Zhang X, Zhang Y, Guo X (2018) Copper oxide nanoparticles for electrochemical detection of organophosphate pesticides. Electrochim Acta 280:1096–1103

53. Zhang X et al (2019) Copper-graphene hybrid material-based electrochemical sensor for the rapid detection of carbamate pesticides. Sens Actuators B: Chem 299:126960
54. Zhang H et al (2020) Electrochemical sensing of organophosphate pesticides using copper nanoparticle-based sensors. Sens Actuators B: Chem 319:128363
55. Zhang et al (2022) Copper nanoparticles/graphene composites for enhanced detection of pesticides. J Agric Food Chem 70(15):4578–4586
56. Zhang C, Qiu M, Wang J, Liu Y (2023) Recent advances in nanoparticle-based optical sensors for detection of pesticide residues in soil. Biosensors 13(4):415
57. Zhang M et al (2023) Electrochemical impedance spectroscopy with copper nanowires for high-sensitivity detection of organophosphate insecticides. Electrochem commun 139:106379
58. Zeng L, Cui H, Chao J, Huang K, Wang X, Zhou Y, Jing T. Colorimetric determination of tetrabromobisphenol A based on enzyme-mimicking activity and molecular recognition of metal-organic framework-based molecularly imprinted polymers. Mikrochim Acta 187:1–9
59. Zheng J et al (2020) Copper nanoparticles as effective sensors for pesticide detection. Environ Sci Technol 54(12):7452–7460
60. Zhou J et al (2017) Copper nanoparticle-based electrochemical sensor for organophosphorus pesticide detection. Sens Actuators B: Chem 249:131–138
61. Zhou L et al (2021) Surface plasmon resonance sensing using copper nanostructures for detecting malathion and diazinon. J Nanopart Res 23(3):65–72
62. Zhou J, Liu Y, Du X, Gui Y, He J, Xie F, Cai J (2023) Recent advances in design and application of nanomaterials-based colorimetric biosensors for agri-food safety analysis. ACS Omega 8(49):46346–46361
63. Zhu H, Shi M, Li S (2017) Colorimetric sensor based on copper nanoparticles for pesticide detection. Sens Actuators B: Chem 242:527–533

Applications of Copper-Based Nanomaterials in Crop Production

Shweta Sharma, Kundan Kumar Chaubey, Parul Singh, Anish Kumar Pal, Faizanul Haque Nagrami, Ashok Kumar Bhatia, Rajnish Kumar, Ashwani Kumar Sanghi, and Kalash Aggarwal

Abstract Copper nanomaterials (CuNMs) can provide a viable alternative to traditional copper bactericides, potentially lowering the growth process of resistance in bacteria. Consequently, this will decrease the rate at which copper builds up in soil and water, reducing the negative effects of copper application on the environment and human health. CuNMs are highly significant and extensively manufactured metal nanomaterials. Their superior conductivity sets them apart from other more stable metal nanomaterials. Moreover, copper is an essential micronutrient that is involved in a number of vital plant functions, such as protein metabolism, enzyme activity, photosynthesis, cell wall construction, mitochondrial respiration, and the reaction to oxidative stress. Although CuNMs have been created via chemical and physical processes, their potential as plant bactericides has yet to be investigated. As a crucial micronutrient for crop growth, CuNMs combined with agrochemicals as herbicides and fertilizers are gaining traction for agricultural use as alternatives to conventional forms. Also, they have been employed in agriculture as an important component of fungicides for combating plant diseases. Copper-based nanoparticles, developed recently, present new opportunities for crop protection with improved efficacy and lower toxicity when compared to bulk Cu formulations. This chapter provides an overview on applications of copper-based nanomaterials in crop production.

S. Sharma · P. Singh · F. H. Nagrami
College of Biotechnology, DUVASU, Mathura, Uttar Pradesh, India

K. K. Chaubey (✉)
Department of Biotechnology, School of Basic and Applied Sciences, Sanskriti University, Mathura, India
e-mail: kundan2006chaubey@gmail.com

A. K. Pal
Division of Research and Innovation, School of Applied and Life Sciences, Uttaranchal University, Dehradun, India

A. K. Bhatia
Department of Microbiology and Immunology, DUVASU, Mathura, Uttar Pradesh, India

R. Kumar · A. K. Sanghi · K. Aggarwal
School of Allied and Health Sciences, MVN University, Palwal, Haryana, India

R. K. Bachheti et al. (eds.), *Copper-Based Nanomaterial for Agricultural Practices*,
Smart Nanomaterials Technology, https://doi.org/10.1007/978-981-96-8610-0_5

Keywords Copper · Nanomaterials · Crop · Production · Agrochemicals

1 Introduction

Copper nanomaterials are defined as particles of at least 50% of the diameters smaller than 100 nm by the European Commission [44]. Their catalytic (mainly), optical, and conductive capabilities led to their early applications in energy conversion and storage, as well as chemical and material manufacture [1, 25, 38]. In recent decades, five primary approaches have been developed to create Cu-based NMs: thermal treatment, sonochemical and photochemical treatments, chemical treatment, and electrochemical synthesis [32, 39, 56, 65]. Of these, this one is the most favored. Biological channels have recently been used to provide more cost-effective, hazardous, waste-free, ecological, and economical methods [17]. According to Rai et al. [45], biogenic CuNMs are a unique family of antimicrobials that show promise against a variety of illnesses, including those that have developed antibiotic resistance [18, 57]. Nonetheless, several unanswered questions remain regarding the specific mode of action of CuNMs. According to data and research, NMs based on copper can alter the permeability of the cellular membrane, leading to cellular death [15, 63]. High concentrations of reactive oxygen species (ROS) lead to oxidative cell damage by oxidizing protein, peroxiding lipids, and destroying DNA [15, 39]. Adding their metallic ions, NMs can also interfere with vital biological functions like DNA replication [63]. Recent research, however, suggests that the CuNMs participate in biocidal action via various distinct mechanisms that cause chemical and physical degradation and downregulation of gene expression [17]. CuNMs have demonstrated effective suppression of alternate, *Penicillium digitatum*, Fusarium sp., *Penicillium italicum*, *Fusarium oxysporum*, *Phomacto niasolani*, and *Curvularia lunata* when it comes to plant protection [42]. Additionally, it has been shown that *Phytophthorain festans* can be suppressed by various sizes of copper nanomaterials and their chemical forms (Cu/Cu_2O, CuO, and Cu_2O) in smaller amounts (15–35 g/hour liter) than the authorized Cu-originated products (35–224 g/hl) [26]. CuNMs have a biocidal effect on bacterium phytopathogens such as *Ralstonia solanacearum*, Xanthomonas spp., and Pseudomonas spp. However, few researches have demonstrated this [11]. However, the bulk of research has focused on how successfully CuNMs combat human diseases like *E. Coli*, *P. aeruginosa*, and *Klebsiella pneumonia* [62].

2 Copper Nanomaterials as a Crop Protective Agent

Extreme weather events are occurring more frequently, posing previously unheard-of difficulties for global agriculture, such as rising food demand and falling crop yields. The application of nanotechnology holds great promise for increasing agriculture yield, improving crop resilience, and reducing pollution in the environment. In this chapter, we provide an overview of the latest research on the implementation of novel nanotechnology to crop production, which may aid in our ability to address agricultural issues [36]. Nanotechnology has numerous applications in plant development and agricultural output using nanoparticles as carriers, including nanopesticides, nanofertilizers, nanosensors, and nanobiotechnology. The unique structure of nanomaterials in agrochemicals, which includes a concentrated size distribution, high specific area for surface coverage, and excellent compatibility with biological applications, facilitates their efficacy and longevity. Additionally, plants can flourish and acquire greater resilience to adversity by applying the appropriate nanomaterials during critical stages of their development or in high-stress situations. Furthermore, new developments in nanobiotechnology and nanotools offer a platform for molecular crop monitoring and modification [36].

Traditional agriculture is experiencing previously unheard-of difficulties as a result of the environment's degradation and the population's rapid rise [54]. For many years, fertilizers and insecticides have been essential to meeting food production demands [28]. However, overuse of these pesticides also prevents agriculture from developing sustainably [48, 58]. Chemical fertilizers are essential to contemporary farming practices. However, for decades, the effectiveness of synthetic chemical products has been declining along with soil, water, and greenhouse gas emissions [54]. Sustainable agriculture may be made possible by nanotechnology. The two types of mineral nutrients that make up nanofertilizers are macronutrients and micronutrients, and they are produced mainly by encasing them in nanomaterials [28]. Macroscopic nutrients have been encapsulated using a variety of nanomaterials such as C, N, K, P, Ca, S, and Mg in order to enhance crop uptake of fertilizers and reduce fertilizer outflow [5, 58]. NP fertilizer composites benefit from increased release efficiency due to NMs' large specific surface area, high stability, and great biocompatibility [68]. For example, there is great potential for urea-hydroxyapatite (HA) nanomaterials to reduce the need for nitrogen fertilizers and lengthen their release duration. The advantages of nanoparticles are derived from the interaction between urea and the amine and carbonyl groups of hydroxyapatites (HA) nanomaterials. In bacterial activity assays, it was demonstrated that Cu nanoparticles inhibited the growth of five distinct bacteria: *Pseudomonas savastanoi* pv. *Erwin iaamylovora, Dickey adadantii, Agrobacterium tumefaciens*, and *Pectobacterium carotovorum* [62]. Furthermore, metallic nanomaterials can inhibit bacteria and fungi while nonmetallic nanoparticles can prevent plant diseases. Chitosan NMs effectively decreased *Xanthomonas campestris* infection in chili peppers compared to the untreated group [23]. Cu nanoparticles and chitosan-polyvinyl alcohol (Cs-PVA) hydrogels applied together improved the expression of SOD in tomatoes under salt stress compared to the control [30].

Numerous horticultural, ornamental, agronomic, and forest plant species are susceptible to a number of illnesses brought on by pathogenic fungus. The most widely used strategy for controlling plant diseases is the use of fungicides targeted to specific plant sites [40]. However, the target fungi can develop distinct resistances to the pesticides, this means that fungi that cause plant diseases can mutate in response to fungicides, losing the fungicide's effectiveness. Copper ions have been used as fertilizers, fungicides, and insecticides for many years. For instance, five copper salts have been applied to lessen grape mildew. There are five drawbacks to using copper ions, like formulations with high copper salt concentrations and instability in field circumstances due to their high solubility in water. Consequently, it is essential to create long-term. Recently, nanotechnology has proposed a number of tools for addressing agricultural difficulties [6–29]. According to reports, developing nanofertilizers and nanotreatments for plant illnesses could increase crop yields through nanotechnology [59]. Many nanomaterials (NMs) have been investigated for the treatment of plant diseases. In particular, copper (Cu) nanomaterials may be effective fungicides. CuNMs should be synthesized using a method that is economical for agricultural use. Numerous synthesis techniques exist, including chemical reduction, hydrothermal, microwave, photochemical, electrochemical, microemulsion, and photochemistry [43].

However, most of them use costly equipment, hazardous chemicals, organic solvents, and crucial reaction conditions. In this regard, a green synthesis technique that makes use of reasonably priced and ecologically acceptable chemicals ought to be suggested for the synthesis of stable Cu nanomaterials. In this instance, we suggest a synthetic technique that uses ascorbic acid to reduce Cu^{2+} ions. Furthermore, ascorbic acid was employed as a stabilizing agent; the food business makes extensive use of these substances. To the greatest extent of our understanding, the two compounds have yet to be combined to create Cu-nanomaterials that are soluble in water. However, in other instances, the only reducing agent was ascorbic acid, while the stabilizing agents were citrate and poly (vinyl alcohol) [34]. Nevertheless, the powders that were described showed signs of significant agglomeration or micro-sized particles, making them unsuitable for use as fungicides. Various fungi, including *Fusarium solani*, *Fusarium oxysporum*, and Neofusicoccum sp. *F. oxysporum* and *F. solani*, affect forest species and crops worldwide, causing numerous ecological and economic issues. These fungi infiltrate the vascular tissues of plants, cause vessel clogging to obstruct water transfer through xylem, and cause wilting of the foliage [64]. Furthermore, a wide variety of distinct perennial fruit crops and tree species have been linked to dieback and canker symptoms due to the fungus Neofusicoccum sp. of the Botryosphaeriaceae family. Research is still lacking against pathogenic fungi, which require examination of Cu-nanomaterials against fungal species that damage species of crops and forests. Recent studies show that Cu-nanomaterials may possess antifungal properties and water stability against *F. oxysporum, F. solani,* and *Neofosicoccum* [42].

CuNMs were successfully created using a green technique involving sodium citrate as a protecting agent and ascorbic acid to chemically break down the copper ions. These green-synthesized CuNM showed strong antifungal efficacy against *F. oxysporum* and Neofusicoccum sp. SEM revealed that, depending on the fungus and compound concentration, CuNMs might cause three distinct types of oxidative stress. The growth inhibition mechanism of CuNM-produced green was clarified using fluorescence microscopy. Inhibition was shown to be due to intracellular ROS production and cell membrane damage. This work proposes a simple and cheap green synthesis method for the preparation of highly effective CuNMs against fungi. Thus, Pariona et al. [37], this nanomaterial can be helpful in the control of dangerous fungi that threaten forest and agricultural species worldwide. Recent applications of copper-based nanoparticles in agriculture have increased yields and shown promising physiological efficacy. Copper concentrations in the environment can be reduced by replacing traditional bulk copper products with these nanoparticles. According to recent research, watermelon plant productivity can be greatly increased, and fungal infections can be considerably reduced by applying commercially available copper oxide and tailored copper nanosheets at leaf concentrations of 100.0–250.0 and 10.0–25.0 ppm, respectively [14].

Before investigating the physiological and molecular activity of the plant, [20] planted alfalfa in a potting mix containing bulk, nano, and ionic copper compounds. They discovered that plants treated with bulk and nano-copper showed improved agronomic response. Sulfur and phosphorus concentrations were lower in plants exposed to bulk and ionic copper than in controls. Overall, the findings suggest that copper nanoparticles may be used as a nanofertilizer and enhance alfalfa physiology. The beneficial application of Cu-based nanoparticles to enhance plant development has been demonstrated by numerous research studies [51, 61].

3 Synthesis of Copper-Based Nanomaterials

Chemical and green synthesis approaches have created Cu-based NMs [10, 21, 31, 46, 55]. Due to its minor environmental impact, the green synthesis of Cu-based NMs utilizing microbes and plants is safer than chemical preparation. Many research institutions have created Cu-based NMs and investigated their antifungal and antibacterial effects against plant diseases. For example, Rubina et al. [49] used the metal vapor synthesis (MVS) method to impregnate chitosan with a CuNM solution to evaluate the antifungal activity. Cu Chitin NCs, which are nanocomposites made of copper nanoparticles, showed inhibitory activity against *Rhizoctonia solani* AG-4 and *Sclerotium rolfsii*, two plant pathogenic fungi. By removing the hyphae and solidifying the cytoplasm, fungal growth was prevented. Saharan et al. [50] prepared ku-chitosan NMs by cross-linking chitosan with tripolyphosphate (TPP) ions using an ion-gelation procedure. They are investigating how well they inhibit the bacteria *Fusarium oxysporum* and *Alternaria solani* to promote tomato plants. Because both fungi inhibited spore germination and mycelia development, the severity of the disease was significantly reduced in both cases. For the relevant fungal infections, Cu-chitosan NMs at doses of 0.10 and 0.12% worked just as well as conventional fungicides. A chitosan-copper (CS–Cu) nanocomposite, which was created without chemical agents, showed remarkable antibacterial capabilities against five strains of both gram-positive and gram-negative bacteria as well as one fungus at 100 μg in another investigation [3]. CuNMs (50–60 nm) produced with gelatin stabilizers significantly decreased the gram-ve bacteria *E. Coli* as well as the gram +ve bacteria *Staphylococcus aureus* and *Bacillus subtilis*. They were discovered to demonstrate bactericidal activities at 7.5 μg/ml and to restrict bacterial cell growth at 3.0 μg/ml. CuNMs hindered cell division, resulting in cell filamentation, which caused cell death and a 96.5% dead cell population, according to [16]. CuNMs encapsulated with Cetyl Tri-methyl Ammonium Bromide (CTAB), with a 3–10 nm particle size, showed strong antifungal efficacy against four fungus in a chemical reduction approach.

4 Copper Oxide-Based Nanosheets

Solanaceous crop bacterial pathogens that cause wilt were tested for CuO nanomaterials antibacterial qualities. The disc diffusion approach examined two bacterial strains: the gram +ve *Clavibacter michiganensis* and the gram–ve *Ralstonia solanacearum.* Against bacterial strains, the synthesized CuO nanosheets exhibited notable inhibitory zones. Diverse metabolic pathways have been observed to be disrupted by Cu ions [35]. The amine and carboxyl groups in the bacterial cell wall demonstrate an excellent affinity for the copper ions [12, 22]. DNA's helical structure may be destroyed by the CuO's binding with bacterial nucleic acid, which has a high surface-to-volume ratio. When the negatively charged component of the bacterial cell wall is attached to the cationic Cu released by nanomaterials, protein denaturation and subsequent cell death can occur [4]. The fabric industry uses methylene blue (MB), a cationic aromatic dye. It affects animals, people, and aquatic life negatively in a number of ways. Consequently, MB and other dangerous colors must be eliminated from industrial effluents. When it comes to different coloring pigments, metal oxide nanomaterials have noteworthy photocatalytic activity [66]. Direct sunlight was used to study the photocatalytic activity of synthesized CuO NS against MB. Following treatment with CuO NS, MB's absorbance band intensity appeared to diminish in a time-dependent manner. Just 4.1% of the dye in a blank degradation experiment showed signs of self-degradation after 140 min. After 140 min of treatment, the catalysts (CuO NS) demonstrated significant MB degradation (82.7%). Recent studies have clearly indicated a gradual rise in the degradation efficiency up to 100 min into the reaction. After this, there was very little MB degradation, which could have been brought on by the lack of reactive sites in CuO NS. Mali and associates have demonstrated up to 90% degradation of MB using Cu nanoparticles made by *Celastrus paniculatus* [41]. The nanosheets show a remarkable bactericidal capacity against bacterial infections that cause wilt in solanaceous plants. The Cu produced as a byproduct in the field may be used by the crops as a micronutrient [13].

5 CuO Nanomaterial-Embedded Hydrogels as Antimicrobial Agent

Recent studies showed the implementation of Cu-based nanomaterials against pathogenic fungal species that cause wilt disease in some plants. This application was demonstrated by a study that evaluated the antibacterial ability of hydrogels embedded with CuO NM by measuring the dehydrogenase activity of *Fusarium oxysporum* and *Fusarium lactucae* after these Cu-based nanoparticles were exposed in vitro. Plants may partially balance their water content in shoots by using hydrogels embedded with CuO NM to decrease the development of abscisic acid in the tissues. This would increase plant resistance to pathogens by triggering the SA-mediated defense response [53]. Overall, hydrogels and Cu treatments considerably raised the SA content in shoots, the Kocide 3000 treatment produced the greatest values. In contrast to Kocide 3000 and CuO NMs, hydrogels embedded with CuO NM performed better in controlling shoot JA levels and root ABA levels. According to this research, hydrogels embedded with CuO NM facilitated communication between SA, JA, and ABA; controlled the response of plants against infection caused by *Fusarium oxysporum* and *Fusarium lactucae*; and successfully boosted the defense response, all of which contributed to the promotion of plant growth [53]. Thus, CuO NM-embedded hydrogels can provide an effective measure to enhance crop productivity.

6 CuO Nanomaterials

Zhao et al. [67] predicted that by 2025, the amount of CuO NMs applied would have increased from 570 tons in 2014 to 1600 tons. CuNMs' well-known antibacterial qualities increase their agricultural efficacy in defending crops against illnesses and plant infections [27, 37, 47, 55]. Half of the world market is made up of the 79,000 tons of Aq. $CuCO_3$ which contains wood preservatives that are manufactured in North America alone. According to reports, Cu nanoparticles with a size range as tiny as 20 nm are present in these wood preservatives [2, 24, 52]. Cu-based NMs have a wide range of highly advantageous agricultural uses as a result, from monitoring environmental contamination to delivering pesticides. However, extended and ongoing use in agriculture may expose the agro-ecosystem to many unknown dangers, particularly about the agricultural soil system. This encompasses the number, diversity, and community of soil microorganisms as well as how they impact other soil processes, such as microbially mediated soil cycles. Their impacts on plants, interactions with the rhizosphere, effects on the food chain, and potential for trophic transfer further cause concerns. Finally, their accumulation in the soil and their interactions with other soil elements, like pollutants, may affect the overall health and quality of the soil system [10] (Figs. 1 and 2).

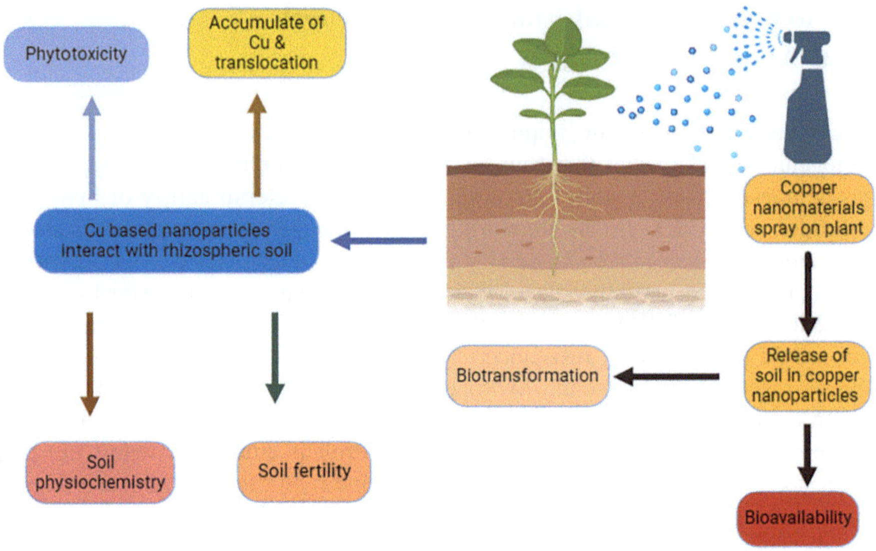

Fig. 1 Effect of copper nanomaterials increase plant growth and rhizospheric soil fertility

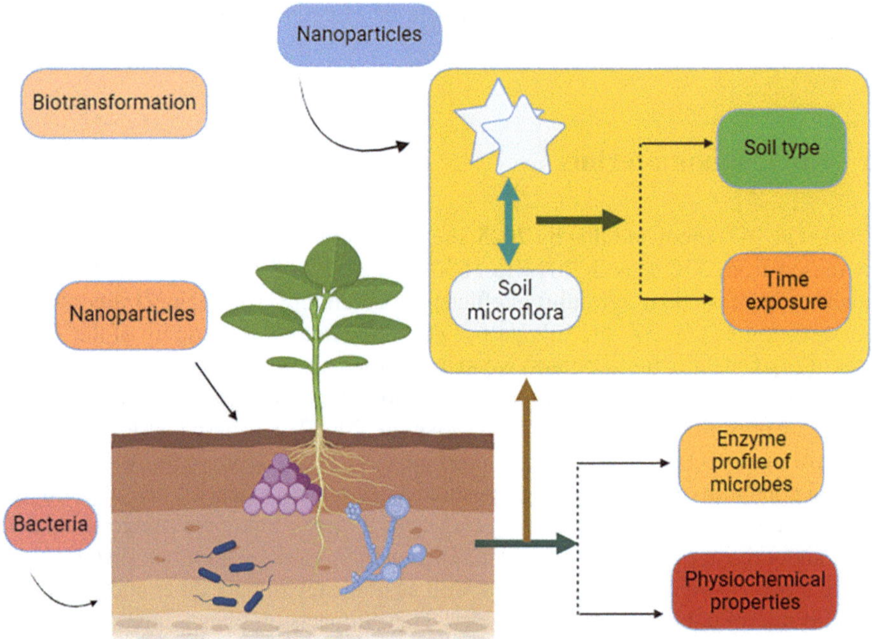

Fig. 2 Effect of copper nanomaterials of microbes and fungi increase the plant growth and soil microflora

7 Copper Nanomaterials as Nanopesticides

The active component of the commercial insecticide Kocide 3000 is copper hydroxide nanomaterials. It is commonly recognized that copper ions have antifungal and antibacterial qualities. Copper nanoparticles have shown antibacterial action against bacterial diseases *Escherichia coli* and *Bacillus subtilis* [19] as well as plant fungal disorders *C. lunata, A. alternata, F. oxysporum, and P. destructiva* [33]. Copper nanoparticles were more effective against bacteria in a study comparing the two than the conventional fungicide bavistin. Additionally, [14] evaluated the effectiveness of copper (II) oxide nanomaterials and copper (II) phosphate trihydrate nanosheets in reducing the *Fusarium oxysporum* f. sp. niveum-caused root fungal disease in watermelon.

8 Zero-Valent Copper Nanomaterials

The roles of the nano-Cuo, which regulate maize plant growth and development and affect the plant's reaction to drought stress, have been well-documented in recent studies. Nano-Cuo has a favorable effect on plant biomass when used under normal and dry conditions. Additionally, the application of nano-Cuo during drought stress preserves the amounts of carotenoid and chlorophyll as well as the hydration state of leaves. Because nano-Cu priming increases the activity of ROS-scavenging enzymes as well as antioxidants, it plays a significant role in regulating the amount of ROS within maize when subjected to water stress. When combined, nano-Cu has the power to favorably regulate a number of defense systems linked to drought tolerance. Most significantly, maize productivity can continue during a drought, which may have applications in sustainable agriculture due to nano-Cu utilization [60].

9 Conclusions and Future Perspectives

The distinct characteristics of copper-based nanomaterials encourage the researchers to consider their application in crop production and agriculture research. Copper-based nanomaterials have established their value in both the agricultural and industrial sectors. They have been shown to increase crop output. Many researchers and scientists have shown interest in utilizing copper-based nanomaterials for crop production in the agricultural sector. Copper-based nanomaterials can be a substitute in the agricultural and industrial sectors for crop production and crop protection from detrimental and non-biodegradable chemicals. The existing investigation study suggests that copper-based nanomaterials have prospective characteristics such as being adaptable, safe, and environment-friendly. So, these copper-based nanomaterials have a wide range of applications in crop production.

The use of copper-based nanomaterials will specifically benefit the improvement of crop production along with crop protection. Copper-based nanomaterials are also recommended as a cost-effective alternative to expensive chemicals in the agriculture field and agriculture industry. Therefore, using copper-based nanomaterials for crop production and crop development will be highly beneficial for the future of the agriculture field and agriculture industry.

Competing Interests The authors declare that they have no competing interests.

Declaration of Generative AI and AI-Assisted Technologies in the Writing Process During the preparation of this work the book chapter author(s) have not taken any assistance from AI, and the book chapter author(s) reviewed and edited the content as needed and take(s) full responsibility for the content of the publication. All authors read and approved the final version of the manuscript.

References

1. Al-Hakkani MF (2020) Biogenic copper nanomaterials and their applications. A review. SN applsci 2(3):505
2. Anjum NA, Adam V, Kizek R, Duarte AC, Pereira E, Iqbal M, Lukatkin AS, Ahmad I (2015) Nanoscale copper in the soil-plant system-toxicity and underlying potential mechanisms. Environ Res 138:306–325
3. Arjunan N, Singaravelu CM, Kulanthaivel J, Kandasamy J (2017) A potential photocatalytic, antimicrobial and anticancer activity of chitosan-copper nanocomposite. Int J Biol Macromol 104:1774–1782
4. Azam A, Ahmed AS, Oves M, Khan MS, Habib SS, Memic A (2012) Antimicrobial activity of metal oxide nanomaterials against Gram-positive and Gram-negative bacteria: a comparative study. Int J Nanomed 7:6003
5. Azeez L, Adejumo AL, Simiat OM, Lateefa (2020) Influence of calcium nanomaterials (canms) on nutritional qualities, radical scavenging attributes of Moringaoleifera and risk assessments on human health. J Food Meascharact 14:2185–2195
6. Bachheti RK, Bachheti A, Husen A (eds) (2023) Nanomaterials for environmental and agricultural sectors. Springer, Singapore. https://doi.org/10.1007/978-981-99-2874-3
7. Bachheti RK, Bachheti A (eds) (2023) Secondary metabolites from medicinal plants: nanoparticles synthesis and their applications. CRC Press, Boca Raton, USA. https://doi.org/10.1201/9781003213727
8. Bachheti A, Bachheti RK, Husen A (eds) (2024) Carbon-based nanomaterials: synthesis, agricultural, biomedical, and environmental interventions. Springer Nature Singapore. https://doi.org/10.1007/978-981-97-0240-4
9. Bachheti RK, Bachheti A, Husen A (eds) (2024) Metal and metal-oxide based nanomaterials: singapore: synthesis, agricultural, biomedical and environmental interventions. Springer Nature Singapore. https://doi.org/10.1007/978-981-99-7673-7
10. Bakshi M, Kumar A (2021) Copper-based nanomaterials in the soil-plant environment: assessing their applications, interactions, fate and toxicity. Chem 281:130940
11. Banik S, Pérez-de-luque A (2017) In vitro effects of copper nanomaterials on plant pathogens, beneficial microbes and crop plants. Span J Agric Res 15
12. Beveridge T, Murray R (1980) Sites of metal deposition in the cell wall of *Bacillus subtilis*. J Bacteriol 141:876–887

13. Bibi H, Iqbal M, Wahab H, Ozturk M, Ke F, Iqbal Z, Khan MI, Alghanem SM (2021) Author correction: green synthesis of multifunctional carbon coated copper oxide nanosheets and their photocatalytic and antibacterial activities. Sci Rep:11(1)16838
14. Borgatta J, Ma C, Hudson-Smith N, Elmer W, Plaza Perez CD, De La Torre-Roche R, Zuverza-Mena N, Haynes CL, White JC, Hamers RJ (2018) Copper based nanomaterials suppress root fungal disease in watermelon (Citrulluslanatus): role of particle morphology, composition and dissolution behavior. ACS Sustain Chemeng 6:14847–14856
15. Chatterjee AK, Chakraborty R, Basu T (2014) Mechanism of antibacterial activity of copper nanomaterials. Nanotechnol 25
16. Chatterjee AK, Sarkar RK, Chattopadhyay AP, Aich P, Chakraborty R, Basu T (2012) A simple robust method for synthesis of metallic copper nanomaterials of high antibacterial potency against *E. Coli*. Nanotechnol 23:085103
17. Chen HF, Wu JJ, Wu MY, Jia H (2019) Preparation and antibacterial activities of copper nanomaterials encapsulated by carbon. New Carbon Mater 34(4):382–389
18. Chowdhury S, Basu A, Kundu S (2014) Green synthesis of protein capped silver nanomaterials from phytopathogenic fungus *Macrophominaphaseolina* (Tassi) Goid with antimicrobial properties against multidrug-resistant bacteria. Nanoscale Res Lett 9:1–1
19. Chung H, Son Y, Yoon TK, Kim S, Kim W (2011) The effect of multi-walled carbon nanotubes on soil microbial activity. Ecotoxicol Environ Saf 74:569–575
20. Cota-Ruiz K, Ye Y, Valdes C, Deng C, Wang Y, Hernández-Viezcas JA, Duarte-Gardea M, Gardea-Torresdey JL (2020) Copper nanowires as nanofertilizers for alfalfa plants: understanding nano-bio systems interactions from microbial genomics, plant molecular responses and spectroscopic studies. Sci Total Environ 742:140572
21. Din MI, Rehan R (2017) Synthesis, characterization, and applications of copper nanomaterials. Anal Lett 50:50–62
22. Doyle RJ, Matthews TH, streipsun (1980) Chemical basis for selectivity of metal ions by the Bacillus subtilis cell wall. J Bacteriol 143:471–480
23. Esyanti RR, Farah N, Bajra BD, Nofitasari D, Martien R, Sunardi S, Safitri R (2020) Comparative study of nano-chitosan and synthetic bactericide application on chili pepper (*Capsicum annuum* L.) infected by Xanthomonascampestris. AGRIVITA J Agricsci 42:11
24. Evans P, Matsunaga H, Kiguchi M (2008) Large-scale application of nanotechnology for wood protection. Nat Nanotechnol 3:577
25. Gawande MB, Goswami A, Felpin FX, Asefa T, Huang X, Silva R, Zou X, Zboril R, Varma RS (2016) Cu and Cu-based nanomaterials: synthesis and applications in catalysis. Chem Rev 116(6):3722–3811
26. Giannousi K, Avramidis I, Dendrinou-Samara C (2013) Synthesis, characterization and evaluation of copper based nanomaterials as agrochemicals against Phytophthorainfestans. RSC Adv 3(44):21743–21752
27. Gopalakrishnan V, Muniraj S (2021) Neem flower extract assisted green synthesis of copper nanomaterials–optimisation, characterisation and antibacterial study. Mater Today Proc 36:832–836
28. Guo H, White JC, Wang Z, Xing B (2018) Nano-enabled fertilizers to control the release and use efficiency of nutrients. Curropin Environ Sci Health 6:77–83
29. Azamal H, Bachheti RK, Bachheti A (eds) (2023) Current trends in green nano-emulsions: food, agriculture and biomedical sectors. Springer Nature Singapore, Singapore. https://doi.org/10.1007/978-981-99-5398-1
30. Hernández-Hernández H, Juárez-Maldonado A, Benavides-Mendoza A, Ortega-Ortiz H, Cadenas-Pliego G, Sánchez-Aspeytia D, González-Morales S (2018) Chitosan-PVA and copper nanomaterials improve growth and overexpress the SOD and JA genes in tomato plants under salt stress. Agronomy 8:175
31. Ingle AP, Duran N, Rai M (2014) Bioactivity, mechanism of action, and cytotoxicity of copper-based nanomaterials: a review. Applmicrobiolbiotechnol 98:1001–1009
32. Joshi A, Sharma A, Bachheti RK, Husen A, Mishra VK (2019) Plant-mediated synthesis of copper oxide nanoparticles and their biological applications. In: Nanomaterials and plant potential. Springer, Cham, pp 221–237. https://doi.org/10.1007/978-3-030-05569-1_8

33. Kanhed P, Birla S, Gaikwad S, Gade A, Seabra AB, Rubilar O, Duran N, Rai M (2014) In vitro antifungal efficacy of copper nanomaterials against selected crop pathogenic fungi. Mater Lett 115:13–17
34. Khanna PK, Gaikwad S, Adhyapak PV, Singh N, Marimuthu R (2007) Synthesis and characterization of copper nanomaterials. Mater Lett 61:4711–4714
35. Kim JH, Cho H, Ryu SE, Choi MU (2000) Effects of metal ions on the activity of protein tyrosine phosphatase VHR: highly potent and reversible oxidative inactivation by Cu^{2+} ion. Arch Biochembiophys 382:72–80
36. Liu C, Zhou H, Zhou J (2021) The applications of nanotechnology in crop production. Molecules 26(23):7070
37. Lopez-Lima D, Mtz-Enriquez AI, Carrión G, Basurto-Cereceda S, Pariona N (2021) The bifunctional role of copper nanomaterials in tomato: effective treatment for Fusarium wilt and plant growth promoter. Scihortic 277:109810
38. Magdassi S, Grouchko M, Kamyshny A (2010) Copper nanomaterials for printed electronics: routes towards achieving oxidation stability. Mater 3(9):4626–4638
39. Mahmoodi S, Elmi A, Hallaj-Nezhadi S (2018) Copper nanomaterials as antibacterial agents. J Mol Pharm Org Process Res 6(1):1–7
40. Malcolm GM, Kuldau GA, Gugino BK, Jiménez-gascomdel M (2013) Hidden host plant associations of soilborne fungal pathogens: an ecological perspective. Phytopathology 103(6):538–544
41. Mali SC, Dhaka A, Githala CK and Trivedi R (2020) Green synthesis of copper nanomaterials using celastruspaniculatuswill.d leaf extract and their photocatalytic and antifungal properties. Biotechnol Rep (Amst) 27:e00518
42. Pariona N, Mtz-Enriquez AI, Sánchez-Rangel D, Carrión G, Paraguay-Delgado F, Rosas-Saito G (2019) Green-synthesized copper nanomaterials as a potential antifungal against plant pathogens. RSC Adv 9(33):18835–18843
43. Parveen F, Sannakki B, Mandke MV Pathan HM (2016) Copper nanomaterials: synthesis methods and its light harvesting performance. Sol Energy Mater Sol Cells 144:371–382
44. Potocnik J (2011) Commission recommendation of 18 October 2011 on the definition of nanomaterial. Off J Eur Communities Legis 275:38–40
45. Rai M, Ingle AP, Pandit R, Paralikar P, Shende S, Gupta I, Biswas JK, da Silva SS (2018) Copper and copper nanomaterials: role in management of insect-pests and pathogenic microbes. Nanotechnol Rev 7(4):303–315
46. Rajesh KM, Ajitha B, Reddy YAK, Suneetha Y, Reddy PS (2018) Assisted green synthesis of copper nanomaterials using Syzygiumaromaticum bud extract: physical, optical and antimicrobial properties. Optik 154:593–600
47. Rajeshkumar S, Rinitha G (2018) Nanostructural characterization of antimicrobial and antioxidant copper nanomaterials synthesized using novel Perseaamericanaseeds. Open 3:18–27
48. Raliya R, Saharan V, Dimkpa C, Biswas P (2018) Nanofertilizer for precision and sustainable agriculture: current state and future perspectives. J Agric Food Chem 66:6487–6503
49. Rubina MS, Vasilkov AYU, Naumkin AV, Shtykova EV, Abramchuk SS, Alghuthaymi MA, Abd-Elsalam KA (2017) Synthesis and characterization ofchitosan–copper nanocomposites and their fungicidal activity against two sclerotia forming plant pathogenic fungi. J Nanostructchem 7:249–258
50. Saharan V, Sharma G, Yadav M, Choudhary MK, Sharma SS, Pal A, Raliya R, Biswas P (2015) Synthesis and in vitro antifungal efficacy of Cu–chitosan nanomaterials against pathogenic fungi of tomato. Int J Biolmacromol 75:346–353
51. Saleem H, Zaidi SJ (2020) Recent developments in the application of nanomaterials in agroecosystems. Nanomat 10(12):2411
52. Servin AD, Pagano L, Castillo-Michel H, De la Torre-Roche R, Hawthorne J, Hernandez-Viezcas JA, Loredo-Portales R, Majumdar S, Gardea-Torresday J, Dhankher OP, White JC (2017) Weathering in soil increases nanoparticle cuo bioaccumulation within a terrestrial food chain. Nanotoxicol 11:98–111

53. Shang H, Ma C, Li C, Zhao J, Elmer W, White JC, Xing B (2021) Copper oxide nanoparticle-embedded hydrogels enhance nutrient supply and growth of lettuce (*Lactuca sativa*) infected with *Fusariumoxysporum* f. Sp. *Lactucae*. Environ Scitechnol 55(20):13432–13442
54. Shang Y, Hasan MK, Ahammed GJ, Li M, Yin H, Zhou J (2019) Applications of nanotechnology in plant growth and crop protection: a review. Molecules 24:2558
55. Shende S, Ingle AP, Gade A, Rai M (2015) Green synthesis of copper nanomaterials by Citrus medica Linn. (Idilimbu) juice and its antimicrobial activity. World J Microbiolbiotechnol 31:865–873
56. Siddiqi KS, Husen A (2020) Current status of plant metabolite-based fabrication of copper/copper oxide nanoparticles and their applications: a review. Biomaterials Research 24(11):1–15. https://doi.org/10.1186/s40824-020-00188-1
57. Singh DP, Ojha AK, Srivastava ON (2009) Synthesis of different Cu (OH) 2 and cuo (nanowires, rectangles, seed-, belt-, and sheetlike) nanostructures by simple wet chemical route. J Physchem C 113(9):3409–3418
58. Singh H, Sharma A, Bhardwaj SK, Arya SK, Bhardwaj N, Khatri M (2021) Recent advances in the applications of nano-agrochemicals for sustainable agricultural development. Environ Sci Process Impacts 23:213–239
59. Tolaymat T, Genaidy A, Abdelraheem W, Dionysiou D, Andersen C (2017) The effects of metallic engineered nanomaterials upon plant systems: an analytic examination of scientific evidence. Sci Total Environ 579:93–106
60. Van Nguyen D, Nguyen HM, Le NT, Nguyen KH, Nguyen HT, Le HM, Nguyen AT, Dinh NT, Hoang SA, Van Ha C (2021) Copper nanoparticle application enhances plant growth and grain yield in maize under drought stress conditions. J Plant Growth Regul 1–2
61. Vandevoort AR, Arai Y (2018) Macroscopic observation of soil nitrification kinetics impacted by copper nanomaterials: implications for micronutrient nanofertilizer. Nanomaterials 8:927
62. Varympopi A, Dimopoulou A, Theologidis I, Karamanidou T, kaldelikerou A, Vlachou A, Karfaridis D, Papafotis D, Hatzinikolaou DG, Tsouknidas A, Skandalis N (2020) Bactericides based on copper nanomaterials restrain growth of important plant pathogens. Pathogens 9(12):1024
63. Yadav L, Tripathi RM, Prasad R, Pudake RN Mittal J (2017) Antibacterial activity of Cu nanomaterials against E. Coli, *Staphylococcus aureus* and *Pseudomonas aeruginosa*. Nano Biomed Eng 9:9–14
64. Yan K, Han G, Ren C, Zhao S, Wu X, Bian T (2018) Fusariumsolani infection depressed photosystem performance by inducing foliage wilting in apple seedlings. Front Plant Sci 9:1–10
65. Wassie AT, Ahmed IN, Bachheti A, Husen A, Bachheti RK (2024) Plant extract-assisted green synthesis and characterization of copper nanocomposite. In: Metal and metal-oxide based nanomaterials: synthesis, agricultural, biomedical and environmental interventions, pp 43–66. Springer Nature Singapore, Singapore. https://doi.org/10.1007/978-981-99-7673-7_3
66. Zhao G, Qin N, Pan A, Wu X, Peng C, Ke F, Iqbal M, Ramachandraiah K, Zhu J (2019) Magnetic nanomaterials metal organic framework composites as sustainable environment adsorbents. J Nanomater 145:4358
67. Zhao S, Su X, Wang Y, Yang X, Bi M, He Q, Chen Y (2020) Copper oxide nanomaterials inhibited denitrifying enzymes and electron transport system activities to influence soil denitrification and N_2O emission. Chemosphere 245:125394
68. Zulfiqar F, Navarro M, Ashraf M, Akram NA, Munné-Bosch S (2019) Nanofertilizer use for sustainable agriculture: advantages and limitations. Plant Sci 289:110–270

Role of Copper-Based Nanomaterials in Soil Remediation

Shweta Sharma, Kundan Kumar Chaubey, Parul Singh, Priyambada Kumari, Minerva Sharma, Anish Kumar Pal, Shalini Sharma, Ashwani Kumar Sanghi, and Ashok Kumar Bhatia

Abstract The scientific study, creation, and alterations of materials and structures at the nanoscale is often referred to as "nanotechnology." The agricultural and industrial sectors have recently paid close attention to soil contamination by metals and metalloids. Certain nanomaterials may enhance the germination of plant seeds and the general growth and development of plants. Scientific interest in integrating nanomaterials inside plants for ecological administration is high. However, if the concentration of the nanomaterial is not adequately controlled in industrial and agricultural sectors, toxicity will likely result. Copper-based nanomaterials have demonstrated significant potential in soil and environmental remediation applications, including CO_2 reduction, wastewater treatment, and biosensing. Copper-based nanomaterials have demonstrated significant potential in soil and environmental remediation applications, including CO_2 reduction, wastewater treatment, and biosensing. Because of their non-toxicity, low allergenicity, and biodegradability, copper-based nanomaterials are an efficient and sustainable method of safeguarding crops or soil during soil

S. Sharma · P. Singh · P. Kumari
College of Biotechnology, DUVASU, Mathura, Uttar Pradesh, India

K. K. Chaubey (✉) · M. Sharma
School of Basic and Applied Sciences, Sanskriti University, Mathura, India
e-mail: kundan2006chaubey@gmail.com

A. K. Pal
Division of Research and Innovation, School of Applied and Life Sciences, Uttaranchal University, Dehradun, India

S. Sharma
Department of Veterinary Physiology and Biochemistry, Lala Lajpat Rai University of Veterinary and Animal Sciences, Hisar, Haryana, India

A. K. Sanghi
School of Allied and Health Sciences, MVN University, Palwal, Haryana, India

A. K. Bhatia
Department of Microbiology and Immunology, College of Biotechnology, DUVASU, Mathura, Uttar Pradesh, India

R. K. Bachheti et al. (eds.), *Copper-Based Nanomaterial for Agricultural Practices*, Smart Nanomaterials Technology, https://doi.org/10.1007/978-981-96-8610-0_6

restoration. Scientists and researchers have demonstrated that copper-based nanoparticles may be utilized for soil remediation and to clean up various polluted surroundings, which will ultimately result in a safer and better world for everyone involved in agriculture. This chapter summarizes the information showing, in terms of soil protection, using copper-based nanomaterials and their special agro nano-chemicals in soil remediation is not only safer for soil in the agricultural sector but also more affordable and kinder to the environment.

Keywords Allergenicity · Biodegradability · Biosensing · Wastewater treatment · Soil remediation

1 Introduction

The population growth has caused a rise in the demand for industrial goods, which has led to an upsurge in industrial effluents released into water and soil streams. Furthermore, various human and natural activities contaminate both land and aquatic habitats. According to Mahmoud and Gan [41], these activities discharge both natural and artificial wastewater that have a bad physiological influence on people and their surroundings and major dangers to flora and fauna. Therefore, the development of sustainable techniques and technologies to treat industrial effluents and municipal garbage, which are heavily polluted with organic and inorganic compounds, is urgently needed [35].

Many traditional and financially unfeasible methods, including biological, physical, and chemical processes, have already been employed in recent decades to remediate the environment; however, nanotechnology has created advancement in terms of specially designed nanomaterials; also, has truly solved this issue [1, 6, 7, 58]. Pertaining to their unique chemical and physiological characteristics and the vital application they employed in the elimination of different contaminating pollutants, the non-valent nanoscale-sized nanomaterials, like zinc, copper, nickel, cobalt, and iron, have drawn the interest of a large community of scientist and researchers [30, 61]. When compared to their microscale metal equivalents, these nanoscale materials have higher reactivities toward different pollutants due to their large surface area relative to weight ratios, which is linked to their high contaminant removal efficiency [25]. Copper is an element that is obtained by applying different electrolytic refining processes to its ores [13], such as chalcopyrite, cuprite, chalcocite, covellite, and malachite [43]. From its revelation approximately circa 3500 BC, it has been used for a wide range of purposes, including the creation of coins and weapons as well as alloys, including metals like zinc (brass), tin (bronze), Ni (cupronickel, constantan), Ag (sterling silver), and others used in manufacturing the industrial items [50]. Other than this utilized in the production of couplers are thermally stable and devices with comparable desired electrical properties [9] because of their high thermal conductivity and electro-conductivity [42, 49]. Additionally, as the field of chemistry of nanomaterials grew [17, 24, 53], copper-based nanomaterials CuO, CuI, CuII, and

CuIII—found many uses in several domains, including as plant biology, medicine, environmental remediation, catalyzing electrical, photochemical, and transformation of organic matter [12, 45, 56] and biological studies of plants [36, 38, 47].

Numerous studies have documented the use of zerovalent copper in environmental remediation projects, as well as the effects of different oxidizing and reducing agents, solution pH, the cooperative effect of immobilizing agents and supporting agents, and different sources of energy in activating nano zerovalent Cu to generate such species that are highly reactive that participate in the catalyzing destruction of different contaminants of organic nature or the elimination of heavy metals by reduction reaction [35].

2 Nanoparticle Synthesis

According to Feng et al. [22], there are several ways to create metallic nanoparticles: cathode corrosion, chemical reduction [11], microwave-enabled [27], laser irradiation [54], reverse-micelle [26], electrochemical, microemulsion, ultrasound [35], thermal decomposition [44], and biogenic formation [19], among other techniques. Since the generated copper nanoparticles are easily oxidized when exposed to air, carrying out these reactions in a naturally inert environment is crucial. This can be achieved by adding argon or nitrogen gas to the reaction vessel.

3 Nanoscale Zerovalent Copper (nZVC)

The size of the produced nanoparticles, the substance used to help, the capping material, the thickness of the surface oxide layer, and additional factors all affect how effective nanostructures are as catalysts for restoring environments [48]. Accordingly, the properties and intended use of the resultant nano-catalyst are greatly influenced by the synthetic method used to create the nanoparticles [32]. Three main groups can be utilized to describe the chemical, biological, and physical mechanisms that lead to nanoparticle production. In order to stabilize nanoparticles and stop copper surface oxidation, protective coatings made of organic and inorganic compounds are widely employed, according to Begletsova et al. [8], Kim et al. [34], Suárez-Cerda et al. [59] (Fig. 1). In order to prepare stable copper nanoparticles, it has been reported that protective agents such as lauric acid, toluol, Dodecyl mercaptan, EtN, C, Si, macrogols (PEG), carbomer, Dodecylbenzene sulfonic acid sodium salt (SDBS)/SDS, are used to prevent oxidative reactions [32].

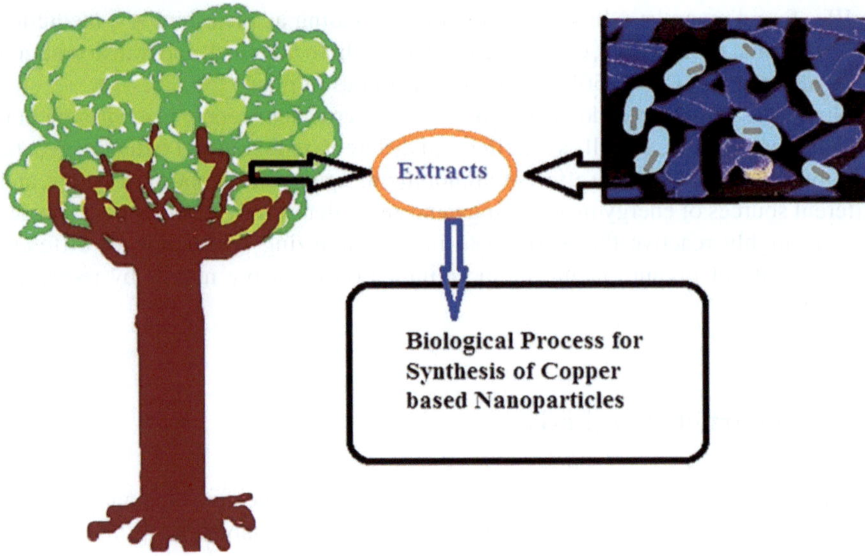

Fig. 1 Biological method for synthesis of Cu nanomaterial

4 Chemical Reduction Method

The most popular technique in literature for generating metallic Cu is the reactive reduction of Aq. Cu salts via agents that are reducing, like ascorbic acid (vitamin C), hydrazine (N_2H_4), sodium hypophosphite ($NaPO_2H_2$), and sodium borohydride ($NaBH_4$). This process yields a consistent particle size distribution, is affordable, simple to understand, executed in an experimental setting, and reproducible. Reduce with $NaBH_4$: Theoretically, $NaBH_4$ must be used with a 4:1 stoichiometric ratio to reduce Cu^{2+}. Cu_2O and $Cu(OH)_2$ were produced as intermediates for the reduction process when the $NaBH_4$ concentration-dependent production of Cu(O) was carried out with lower $NaBH_4$ concentrations; these results were further supported by XRD examination [39]. The signal corresponding to $Cu(OH)_2$ vanished when $NaBH_4$ concentration increased. However, Cu_2O contamination persisted until $NaBH_4$ concentration increased eight times beyond the stoichiometric dosage.

Additionally, it was noted that when the $NaBH_4$ was given, the size of the copper nanoparticles reduced as the reduction process intensified. Reduction by hydrazine: Cu (II) is reduced to Cu(I) at room temperature, while Cu(I) is converted to Cu(O) at higher temperatures. The nitrogen gas generated during the process provides an inert environment for the reduction procedure [4].

According to Kim et al. [33] and Liu et al. [37], the particular structure of nZVC created via an approach, namely bottom-up, is a shell structure with core-zerovalent metallic copper serving and an oxide shell made up of Cu_2O and CuO. Because copper nanoparticles are pyrophoric, surface oxidation is unavoidable unless manufactured and kept in an inert environment [2]. The electron powerhouse is the nZVC core. According to Fathima et al. [21] and Shikha et al. [57], impurities in core–shell nanoparticles' surface are reductively degraded due to the charge transfer from the core through the intermediate semi-conducting copper oxide layer. Stabilizing and capping agents prevent surface oxidation, giving the nZVC additional stability [16]. Kim et al. [33] created Cu-copper formate core–shell nanoparticles to stop the oxidation of nZVC's surface. The synthetic nZVC is more stable due to the organic/inorganic support utilized for immobilization, which decreases aggregation and Cu(I/II) leaching under unfavorable circumstances, Pistachio shell powder [35], zeolite [14], Fe_2O_3, green rust, $ZrSiO_4$, microscale zinc, chitosan, TiO_2, cellulose filter paper, carboxymethyl cellulose, carbon nanotubes, graphene, and montmorillonite. The configuration of the study and the number of adjacent materials employed define the size and surface area of the artificial zerovalent copper nanoparticles.

5 Physical Method

The physical method used to create copper nanoparticles is a top-down strategy in which bulk copper is first reduced in size to the nanoscale by a variety of mechanical techniques, including etching, milling, pulsed ablation with a laser, and noble gas sputtering [35]. Physical methods of copper nanoparticles synthesis are electro-spraying, evaporation–condensation pyrolysis by laser, and ablation by laser. The pulsed laser-induced ablation approach has garnered greater interest among these physical procedures due to its ease of use and environmental friendliness. The physical method produces copper nanoparticles of desired morphological identities for soil remediation [64]. The other advantages of physical methods of copper nanoparticles for soil remediation are cost infectiveness, operation skills, high catalytic performance, reusability, and high purity [3].

6 Mechanism of Contaminant Removal

Reductive breakdown, oxidative elimination, and adsorption onto the surface are the three main removal processes by which the Cu nanoparticles have been involved in the environmental remediation phenomenon, according to studies conducted to determine the different varieties of copper engaged within the elimination of a variety of pollution, both organic and inorganic (De Sousa et al. 2019a). When there is no dissolved oxygen present (anoxic environment), the removal process mostly entails the exchange of electrons from CuO, which leads to the reductive breakdown and elimination of organic and inorganic contaminants. Furthermore, the prevention of the production of hydroxyl radicals guarantees that CuO is stable in the absence of oxygen. Conversely, in the presence of oxygen (oxic), CuO undergoes electron transfer to generate a range of reactive oxygen species (ROS), including H_2O_2, O_2^-, OH, and so on. This leads to the oxidative destruction of organic pollutants [62].

Table 1 Chemical groups studied for removing contaminants

Chemical group (Inhibitor)	Inhibited component	References
Tert-butyl alcohol	OH^-	[51]
SOD	Anions of superoxide nature (O_2^-)	[18]
NaN_3	Singlet oxygen	[23]
Catalase	Hydrogen peroxide	[55]
Isopropanol + TBA	Hydroxyl radical	[40]

A variety of scavengers of radicals have been investigated to identify the important species of radicals that contribute to the degradation pathways.

The pH of the reaction mixture has a major impact on how contaminants are eliminated. Fewer than 7 pH causes the Cu^+ species to predominate in the solution, which encourages the in situ synthesis of H_2O_2 from dissolved O_2 and H_2O in a process similar to that of Fenton. Later on, the Cu^+ species also contributes to the breakdown of H_2O_2 into OH radicals and regulate the process of degradation as a whole [63]. Because copper nanoparticles' surfaces allow copper oxides (CuO) and hydroxides (Cu(OH2)) to form at higher pH values (above 7), surface adsorption is the predominant method of removing pollutants (Table 1) [46]. For the different decline processes, zerovalent copper has also been investigated in conjunction with external oxidizing and reducing agents such as NaBH [52], peroxymonosulfate [65] and H_2O_2 [40].

7 Elimination of Organic Pollutants

Organic pollutants, often known as waste, contain carbon, hydrogen, and other heteroatoms. Depending on their physical makeup, these pollutants can be found in the environment as solids (chemicals, waxes, polymers, resins) or as gases or liquids (volatile organic compounds or VOCs). Their intricate structure makes them very toxic, making them a constant threat to plants and animals. Scientists worldwide are frantically trying to solve this issue by creating novel approaches and cutting edge strategies. Degradation of organic pollutants by tailored nanoparticles offers a long-term, sustainable solution. Unraveling substance mentioned in previous research since the elimination of both inorganic and organic contaminants is zerovalent copper [35].

8 Remediation of Different Elements

The method of treating wastes, such as solid waste, industrial waste, and waste water, is called remediation. Though many and diverse, remediation techniques may be broadly divided into two groups: ex situ and in situ techniques. Ex situ techniques entail the surface treatment of contaminated groundwater, the excavation of impacted soils, and extraction. By treating the pollution in place, in situ techniques aim to prevent the removal of groundwater or soil. To improve the development and metabolism of the soils, extra nutrients such as vitamins, minerals, and pH buffers are supplied in both approaches (Fig. 2). Phytoremediation, bioventing, bio-attenuation, bio-sparging, composting (biopiles and windrows), and land farming are a few examples of components connected to bio-remediation. Other remedial techniques include thermal desorption, vitrification, air stripping, bioleaching, rhizo filtration, and soil washing. Bioremediation aims to eliminate or minimize hazardous substances to enhance the quality of soil and water (Fig. 3) [10, 31]. Different components of remediation include air sparging/bio-sparging, soil vapor extraction, sparge and vent systems, dual-phase extraction systems, low profile air strippers, oil/water separators, liquid and vapor phase carbon adsorption systems, thermal/catalytic oxidizers, and heavy metal treatment processes. The most prevalent anthropogenic terrestrial contaminants pose a serious hazard to the ecosystem since they are poisonous and tenacious. Remediations of such components with different copper-based nanoparticles have been explained in Table 2.

9 Conclusions and Future Perspectives

The diverse characteristics of copper-based nanoscale substances encourage researchers to believe in their application in soil remediation to advance cultivation research and agricultural production. The investigation and research study shows that copper-based nanomaterials have the approaching characteristics for soil remediation. Thus, these nanoparticles with a copper basis have a broad spectrum of applications in soil remediation. The terrestrial contaminants pose a serious danger to the environment since they are lethal and persistent. The use of copper-based nanomaterials promises several options and brings about a new aspect into the agriculture sector by solving the problems of soil remediation. Utilizing nanoparticles based on copper will specifically benefit the improvement of soil repair and crop protection. Improving soil and water quality by eliminating or reducing hazardous chemicals is the ultimate aim of soil remediation. Therefore, applying copper-based nanomaterials in soil remediation will be highly helpful for the future of the agriculture sector.

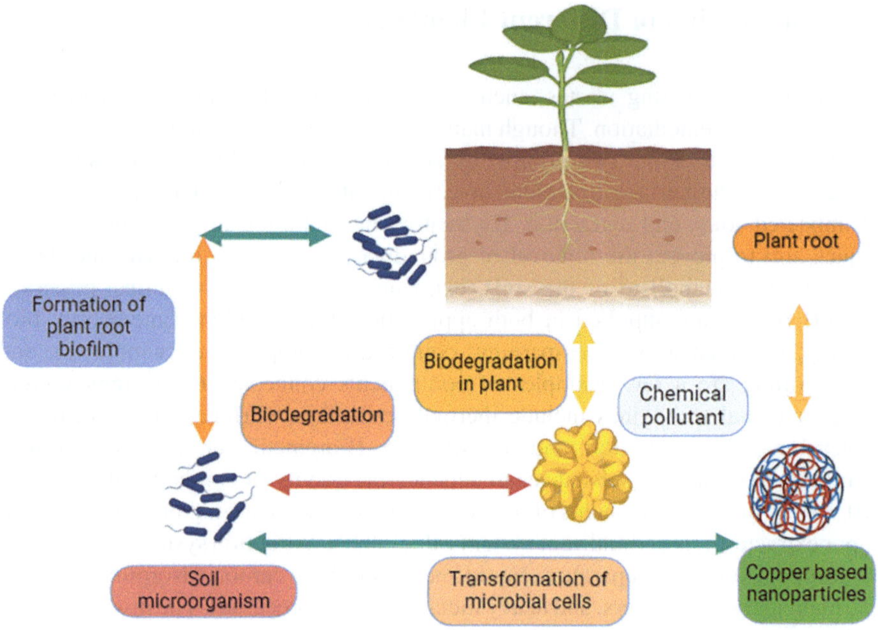

Fig. 2 Copper nanoparticles increase soil fertility and plant root biofilm formation

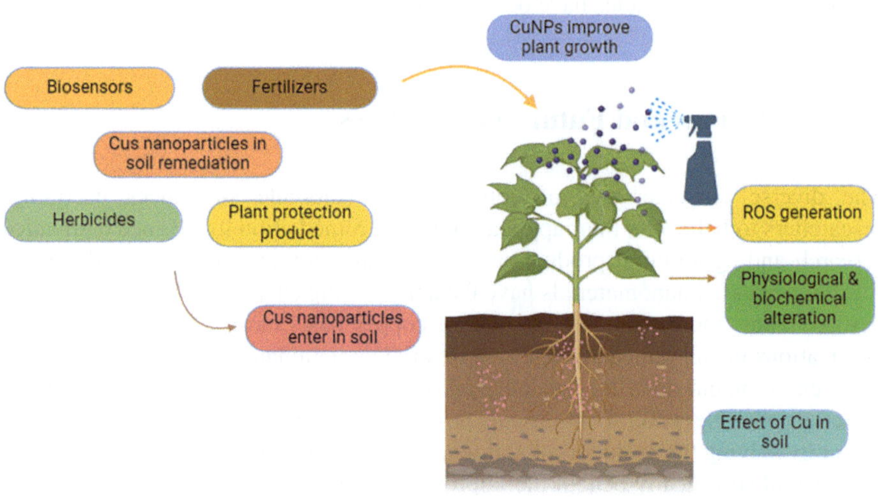

Fig. 3 Copper nanoparticles improve soil and plant protection activity against ROS

Table 2 Remediation of several components by Cu-based nanoparticles

Remediation (Component)	Discharge source	Process	References
Chlorinated organic compounds	• Pesticides • Dry cleaners • Grease removers	Zerovalent copper (ZVC), with its size ~ 50 nm or small in size and high surface area, was shown to be an efficient catalyst for dichloromethane dechlorination processes under NaBH4 reduction conditions	[28]
Oil/Petroleum	Oil production, exploration, and utilization	Suspensions of Cu nanoparticles with Fe nanoparticles and biosurfactants. Process followed was first-order reaction	[60]
Chromated copper arsenate	Contaminated soils polluted with in-situ immobilized arsenic	Fe/Cu nanoparticles stabilized by starch (0 04 wt %) at a soil: liquid ratio = 0 1, was added to the contaminated soil	[5]
1,2-Dichloroethane	PVC industrial plants	Zerovalent copper nanoparticles using NaBH4 as electron donor	[29]
Tetrabromobisphenol A (TBBPA)	Brominated flame retardant	nZVC with Cl-ion in green rust shows enhanced reduction reaction against TBBPA	[20]

Competing Interests The authors declare that they have no competing interests.

Declaration of Generative AI and AI-Assisted Technologies in the Writing Process During the preparation of this work the book chapter author(s) have not taken any assistance from AI, and the book chapter author(s) reviewed and edited the content as needed and take(s) full responsibility for the content of the publication. All authors read and approved the final version of the manuscript.

References

1. Akhtar FZ, Archana KM, Krishnaswamy VG, Rajagopal R (2020) Remediation of heavy metals (Cr, Zn) using physical, chemical and biological methods: a novel approach. SN Appl Sci 2:1–4
2. Alymov M, Rubtsov N, Seplyarskii B, Zelenskii V, Ankudinov A, Tsvetkov G, Chernysh V (2018) Effect of production conditions on the size of copper nanoparticles and the modes of ignition and combustion of a copper nanopowder in air, Doklady. Chem Spring 318–322
3. Amendola V, Meneghetti M (2013) What controls the composition and the structure of nanomaterials generated by laser ablation in liquid solution. Phys Chem Chem Phys 15(9):3027–3046
4. Andal V, Buvaneswari G (2017) Effect of reducing agents in the conversion of Cu_2O nanocolloid to Cu nanocolloid. Eng Sci Tech Int J 20:340–344
5. Babaee Y, Mulligan CN, Rahaman MS (2018) Arsenic immobilization in soil using starch-stabilized Fe/Cu nanoparticles: a case study in treatment of a chromated copper arsenate (CCA)-contaminated soil at lab scale. J Soils Sediments 18:1610–1619

6. Bachheti A, Bachheti RK, Husen A (eds) (2024) Carbon-based nanomaterials: synthesis, agricultural, biomedical, and environmental interventions. Springer Nature Singapore. https://doi.org/10.1007/978-981-97-0240-4
7. Bachheti RK, Bachheti A, Husen A (eds) (2024) Metal and metal-oxide based nanomaterials: singapore: synthesis, agricultural, biomedical and environmental interventions. Springer Nature Singapore. https://doi.org/10.1007/978-981-99-7673-7
8. Begletsova N, Selifonova E, Chumakov A, Al-Alwani A, Zakharevich A, Chernova R, Glukhovskoy E (2018) Chemical synthesis of copper nanoparticles in aqueous solutions in the presence of anionic surfactant sodium dodecyl sulfate
9. Berman R, Macdonald DKC (1952) The thermal and electrical conductivity of copper at low temperatures. Proc Roy Soc Lond Math Phys Sci 211:122–128
10. Canak S, Berezljev L, Borojevic K, Asotic J, Ketin S (2019) Bioremediation and green chemistry. Fresenius Environ Bull 28(4):3056–3064
11. Chang PH, Chou TH, Sahu RS, Shih YH (2019) Chemical reduction-aided zerovalent copper nanoparticles for 2, 4-dichlorophenol removal. Appl Nanosci 9:387–395
12. Chen C, Peters JC, Fu GC (2021) Photoinduced copper-catalysed asymmetric amidation via ligand cooperativity. Nat 596:250–256
13. Daehn K, Allanore A (2020) Electrolytic production of copper from chalcopyrite. Curr Opin Electrochem 110–119
14. Danish M, Gu X, Lu S, Ahmad A, Naqvi M, Farooq U, Zhang X, Fu X, Miao Z, Xue Y (2017) Efficient transformation of trichloroethylene activated through sodium percarbonate using heterogeneous zeolite supported nano zero valent iron-copper bimetallic composite. Chem Eng J 308:396–407
15. De Sousa PVF, de Oliveira AF, da Silva AA, Lopes RP (2019) Environmental remediation processes by zero valence copper: reaction mechanisms. Environ Sci Pollut Control Ser 26:14883–14903
16. Din MI, Rehan R (2017) Synthesis, characterization, and applications of copper nanoparticles. Anal Lett 50:50–62
17. Dong XY, Zhang YF, Ma CL, Gu QS, Wang FL, Li ZL, Jiang SP, Liu XY (2019) A general asymmetric copper-catalysed Sonogashira C (sp 3)–C (sp) coupling. Nat Chem 11:1158–1166
18. Dror I, Fink L, Weiner L, Berkowitz B (2020) Elucidating the catalytic degradation of enrofloxacin by copper oxide nanoparticles through the identification of the reactive oxygen species. Chem 258:127266
19. El-Saadony MT, El-Hack A, Mohamed E, Taha AE, Fouda MM, Ajarem JSN, Maodaa S, Allam AA, Elshaer N (2020) Ecofriendly synthesis and insecticidal application of copper nanoparticles against the storage pest *Tribolium castaneum*. Nanomat 10:587
20. Fang L, Liu R, Xu L, Li J, Huang LZ, Li F (2019) Enhanced debromination of tetrabromobisphoenol a by zero-valent copper-nanoparticle-modified green rusts. Environ Sci J Integr Environ Res NANO 6:970–980
21. Fathima JB, Pugazhendhi A, Oves M, Venis R (2018) Synthesis of eco-friendly copper nanoparticles for augmentation of catalytic degradation of organic dyes. J Mol Liq 260:1–8
22. Feng J, Chen D, Sediq AS, Romeijn S, Tichelaar FD, Jiskoot W, Yang J, Koper MT (2018) Cathodic corrosion of a bulk wire to nonaggregated functional nanocrystals and nanoalloys. ACS Appl Mater Interfaces 10:9532–9540
23. Gao Y, Wu T, Yang C, Ma C, Zhao Z, Wu Z, Cao S, Geng W, Wang Y, Yao Y (2021) Activity trends and mechanisms in peroxymonosulfate-assisted catalytic production of singlet oxygen over atomic metal-N-C catalysts. Angew Chem Int Ed 60:22513–22521
24. Gawande MB, Goswami A, Felpin FX, Asefa T, Huang X, Silva R, Zou X, Zboril R, Varma RS (2016) Cu and Cu-based nanoparticles: synthesis and applications in catalysis. Chem Rev 116:3722–3811
25. Guerra E, Llompart M, Garcia-Jares C (2018) Analysis of dyes in cosmetics: challenges and recent developments. Cosm 5(3):47
26. Gutierrez JA, Silber JJ, Falcone RD, Correa NM (2021) Modified reverse micelle method as facile way to obtain several gold nanoparticle morphologies. J Mol Liq 331:115709

27. Hashimi AS, Ginting RT, Chin SX, Lau KS, Nohan MANM, Zakaria S, Yap CC, Chia CH (2020) Fast microwave-assisted synthesis of copper nanowires as reusable high-performance transparent conductive electrode. Curr Appl Phys 20:205–211
28. Huang CC, Lo SL, Lien HL (2012) Zero-valent copper nanoparticles for effective dechlorination of dichloromethane using sodium borohydride as a reductant. Chem Eng J 203:95–100
29. Huang CC, Lo SL, Tsai SM, Lien HL (2011) Catalytic hydrodechlorination of 1, 2-dichloroethane using copper nanoparticles under reduction conditions of sodium borohydride. J Environ Monit 13:2406–2412
30. Husen A, Bachheti RK, Bachheti A (eds) (2023) Current trends in green nano-emulsions: food, agriculture and biomedical sectors. Springer Nature Singapore, Singapore. https://doi.org/10.1007/978-981-99-5398-1
31. Kensa VM (2011) Bioremediation—an overview. Control Pollut 27(2):161–168
32. Khodashenas B, Ghorbani HR (2014) Synthesis of copper nanoparticles: an overview of the various methods. Kor J Chem Eng 31:1105–1109
33. Kim I, Kim Y, Woo K, Ryu EH, Yon KY, Cao G, Moon J (2013) Synthesis of oxidation-resistant core–shell copper nanoparticles. RSC Adv 3:15169–15177
34. Kim S, Kim JH, Kwon WY, Hwang SH, Cha BS, Kim JM, Oh SS, Park KS (2019) Synthesis of DNA-templated copper nanoparticles with enhanced fluorescence stability for cellular imaging. Microchim Acta 186:1–5
35. Kumar S, Kaur P, Brar RS Babu JN (2022) Nanoscale zerovalent copper (nZVC) catalyzed environmental remediation of organic and inorganic contaminants: a review. Hel 8(8):e10140
36. Lamichhane JR, Osdaghi E, Behlau F, K€ohl J, Jones JB, Aubertot J N (2018) Thirteen decades of antimicrobial copper compounds applied in agriculture a review. Agro Susta Devel 38:1–18
37. Liu C, Dong H, Ji Y, Rujisamphan N, Li Y (2019) High-performance hydrogen evolution reaction catalysis achieved by small core-shell copper nanoparticles. J Colloid Interface Sci 551:130–137
38. Liu J, Dhungana B, Cobb GP (2018) Environmental behavior, potential phytotoxicity and accumulation of copper oxide nanoparticles and arsenic in rice plants. Environ Toxicol Chem 37:11–20
39. Liu QM, Zhou DB, Yamamoto Y, Ichino R, Okido M (2012) Preparation of Cu nanoparticles with NaBH4 by aqueous reduction method. Trans Nonferrous Metals Soc China 22:117–123
40. Ma X, Cheng Y, Ge Y, Wu H, Li Q, Gao N, Deng J (2018) Ultrasound-enhanced nanosized zero-valent copper activation of hydrogen peroxide for the degradation of norfloxacin. Ultrason Sonochem 40:763–772
41. Mahmoud SH, Gan TY (2018) Impact of anthropogenic climate change and human activities on environment and ecosystem services in arid regions. Sci Total Environ 633:1329–1344
42. Minneci RP, Lass EA, Bunn JR, Choo H, Rawn CJ (2021) Copper-based alloys for structural high-heat-flux applications: a review of development, properties, and performance of Cu-rich Cu–Cr–Nb alloys. Int Mater Rev 66:394–425
43. Moskalyk R, Alfantazi A (2003) Review of copper pyrometallurgical practice: today and tomorrow. Miner Eng 16:893–919
44. Numaga N, Hayashi H, Smith RL (2020) Supercritical hydrothermal synthesis of polyacrylic acid-capped copper nanoparticles and their feasibility as conductive nanoinks. J Electron Mater 49:5681–5686
45. Ojha NK, Zyryanov GV, Majee A, Charushin VN, Chupakhin ON, Santra S (2017) Copper nanoparticles as inexpensive and efficient catalyst: a valuable contribution in organic synthesis. Coord Chem Rev 353:1–57
46. Ourique MF, Sousa PV, Oliveira AF, Lopes RP (2018) Comparative study of the direct black removal by Fe, Cu, and Fe/Cu nanoparticles. Environ Sci Pollut Control Ser 25:28928–28941
47. Panagos P, Ballabio C, Lugato E, Jones A, Borrelli P, Scarpa S, Orgiazzi A, Montanarella L (2018) Potential sources of anthropogenic copper inputs to European agricultural soils. Susta 10:2380
48. Pasinszki T, Krebsz M (2020) Synthesis and application of zero-valent iron nanoparticles in water treatment environmental remediation catalysis and their biological effects. Nanomaterials 10:917

49. Pickard SM, Johnson TG, Kuang K. (2021) Development of high thermal conductivity copper alloys with diamond particle additions 23rd European microelectronics and packaging conference and exhibition (EMPC). IEEE, pp 1–6
50. Powell C, Webster P (2019) Copper alloys, the corrosion performance of metals for the marine environment: a basic guide. CRC Press, pp 26–41
51. Qi C, Liu X, Ma J, Lin C, Li X, Zhang H (2016) Activation of peroxymonosulfate by base: implications for the degradation of organic pollutants. Chem 151:280–288
52. Raut SS, Kamble SP, Kulkarni PS (2016) Efficacy of zero-valent copper (CuO) nanoparticles and reducing agents for dechlorination of mono chloroaromatics. Chem 159:359–366
53. Sambiagio C, Marsden SP, Blacker AJ, McGowan PC (2014) Copper catalysed Ullmann type chemistry: from mechanistic aspects to modern development. Chem Soc Rev 43:3525–3550
54. Shahzeydi A, Ghiaci M, Farrokhpour H, Shahvar A, Sun M, Saraji M (2019) Facile and green synthesis of copper nanoparticles loaded on the amorphous carbon nitride for the oxidation of cyclohexane. Chem Eng J 370:1310–1321
55. Shen W, Mu Y, Wang B, Ai Z, Zhang L (2017) Enhanced aerobic degradation of 4-chlorophenol with iron-nickel nanoparticles. Appl Surf Sci 393:316–324
56. Shi SL, Buchwald SL (2015) Copper-catalysed selective hydroamination reactions of alkynes. Nat Chem 7:38–44
57. Shikha J, Ankita J, Kachhawah P, Devra V (2015) Synthesis and size control of copper nanoparticles and their catalytic application. Trans Nonferrous Metals Soc China 25:3995–4000
58. Siddiqi KS, Husen A (2020) Current status of plant metabolite-based fabrication of copper/copper oxide nanoparticles and their applications: a review. Biomater Res 24(11):1–15. https://doi.org/10.1186/s40824-020-00188-1
59. Suárez-Cerda J, Espinoza-Gomez H, Alonso-Nú~nez G, Rivero IA, Gochi-Ponce Y, Flores-Lopez LZ (2017) A green synthesis of copper nanoparticles using nativecyclodextrins as stabilizing agents. J Saudi Chem Soc 21:341–348
60. Vu KA, Mulligan CN (2023) Remediation of oil-contaminated soil using Fe/Cu nanoparticles and biosurfactants. Environ Technol 44(22):3446–3458
61. Wassie AT, Ahmed IN, Bachheti A, Husen A, Bachheti RK (2024) Role of zinc oxide nanomaterials for photocatalytic degradation of environmental pollutants. In: Metal and metal-oxide based nanomaterials: synthesis, agricultural, biomedical and environmental interventions. Springer Nature Singapore, Singapore, pp 287–311. https://doi.org/10.1007/978-981-99-7673-7_14
62. Wen G, Wang SJ, Ma J, Huang TL, Liu ZQ, Zhao L, Xu JL (2014) Oxidative degradation of organic pollutants in aqueous solution using zero valent copper under aerobic atmosphere condition. J Hazard Mater 275:193–199
63. Xu L, Yang Y, Li W, Tao Y, Sui Z, Song S, Yang J (2019) Three-dimensional macroporous graphene-wrapped zero-valent copper nanoparticles as efficient microelectrolysis-promoted Fenton-like catalysts for metronidazole removal. Sci Total Environ 658:219–233
64. Yan Z, Chrisey DB (2012) Pulsed laser ablation in liquid for micro-/nanostructure generation. J Photochem Photobiol 13(3):204–223
65. Zhou P, Liu B, Zhang J, Zhang Y, Zhang G, Wei C, Liang J, Liu Y, Zhang W (2016) Radicals induced from peroxomonosulfate by nanoscale zero-valent copper in the acidic solution. Water Sci Technol 74:1946–1952

Use of Copper-Based Nanomaterials for Nutrient Delivery and Uptake

Durgesh Wadhwa, Kundan Kumar Chaubey, D. S. Tomar, Hridesh Kumar, Samriddhi Singh, Tisha Sharma, Ashwani Kumar Sanghi, Vineet Kumar, and Shailendra Thapliayal

Abstract Copper-based nanomaterials (Cu-based NMs) have led to considerable advancements in improving crop production and sustainability in agriculture. Cu nanoparticles (CuNPs) and Cu oxide nanoparticles (CuONPs), i.e., Cu-based NMs, have unique physicochemical properties for enhancing nutrient management systems, i.e., nutrient delivery and uptake, and they make Cu-based NMs promising applications tools for use in the agriculture sector. Cu-based NMs facilitate nutrient delivery and uptake in plants and improve crop growth and yield in the agriculture sector. Cu-based NMs are highlighted in today's scenario due to their ability to increase the nutrient solubility, nutrient bioavailability, and stability of agricultural crops. Cu-based NMs release essential nutrients, including micronutrients and macronutrients, for plants and soils, thus reducing nutrient deficiencies in plants and soils. These Cu-based NMs diminish nutrient discharge, decrease environmental pollution, and contribute to agricultural sector sustainability.

Keywords Cu-based NMs · CuNPs · Nutrient delivery · Bioavailability · Crop growth · Nutrient uptake

---

D. Wadhwa · H. Kumar · S. Singh
Department of Chemistry, School of Basic and Applied Sciences, Sanskriti University, Mathura, Uttar Pradesh, India

K. K. Chaubey (✉) · D. S. Tomar · T. Sharma
Department of Biotechnology, School of Basic and Applied Sciences, Sanskriti University, Mathura, Uttar Pradesh, India
e-mail: kundan2006chaubey@gmail.com

A. K. Sanghi · V. Kumar
School of Allied and Health Sciences, MVN University, Palwal, Harryana, India

S. Thapliayal
UIM, Uttaranchal University, Dehradun, Uttarakhand, India

R. K. Bachheti et al. (eds.), *Copper-Based Nanomaterial for Agricultural Practices*, Smart Nanomaterials Technology, https://doi.org/10.1007/978-981-96-8610-0_7

1 Introduction

Nutrient delivery and uptake in plants increase the crop quality and production [17]. Nanomaterials (NMs) have promising distinctive physicochemical properties and have a high potential to manage nutrient management in the agricultural industry [4–6, 29]. Using nanotechnologies, we can improve the nutrient delivery system and absorption in soils, and we also improve the physical and chemical properties of NPs in various applications in the agricultural and industrial sectors. Nowadays use of Cu nanoparticles (CuNPs) is increasing day by day due to their unique properties, help in the transportation and absorption of nutrient delivery to the soil and crops to gain more agricultural production [2]. Yadav et al. [76] reported that CuNPs and Cu oxide (CuO) NPs could significantly increase soils' and plants' ability to absorb, transport, and uptake nutrients for their different processes. These NPs work as carriers for transporting vital nutrients to soils and plants and also enhance the ability of plants, roots, and soils to absorb vital nutrients [52]. Copper-based nanomaterials (Cu-based NMs) facilitate various types of activities like the movement of nutrients, enhance the symbiotic relationships between plants and soil fungi, release substances from plant roots, and alleviate nutrient-related stress [39, 73]. The different benefits and difficulties of CuNPs are discussed in Table 2, showing potential future applications of CuNPs for sustainable nutrient delivery and management in the agriculture sector. There are lots of challenges in the nutrient delivery system, like soil degradation, nutrient deficiencies, and environmental stress that hamper the growth of the plants. Nutrient loss due to leaching is also a big challenge that results in pollution in the environment and poor growth of the plants. To overcome those issues, NMs are the only option for delivering nutrients as carriers. Due to the involvement of Cu in photosynthesis, respiration, and protein synthesis, we are considering it as a critical trace element in plants [9]. Cu oxide nanoparticles (CuO NPs), Cu nanoparticles (CuNPs), and Cu sulfide nanoparticles (CuS NPs) are small in size, have high surface area, and tunable properties make them highly promising in nutrient uptake and delivery.

There is a strong interaction demonstrated between plants and Cu-based NMs for potential uptake of nutrients, growth rate, and increased resistance against pathogens and abiotic stresses. CuO NPs enhance the absorption of essential nutrients like nitrogen, potassium, and phosphorus, and improving plant tolerance toward oxidative stress and disease [16]. For example, Cu-based NMs are responsible for carrying the slow-release fertilizers, promoting the maintenance of higher levels of nutrients [8]. The toxicity, as well as environmental impact, also raises the big concern. The size, concentration, surface charge, and exposure duration of Cu-based NMs impact the toxicity level of Cu-based NMs [56]. Cu is the most important micronutrient element for plants, excessive production of Cu in plant tissues causes phytotoxicity that can harm plant growth and the entire food chain of plants [79]. So, a balanced approach to nutrient delivery will be mandatory because increased nutrient uptake does not cause a harmful impact on the environment. In this chapter, we are focusing on Cu-based NMs' role in nutrient delivery and uptake in plants. Here, we also examine

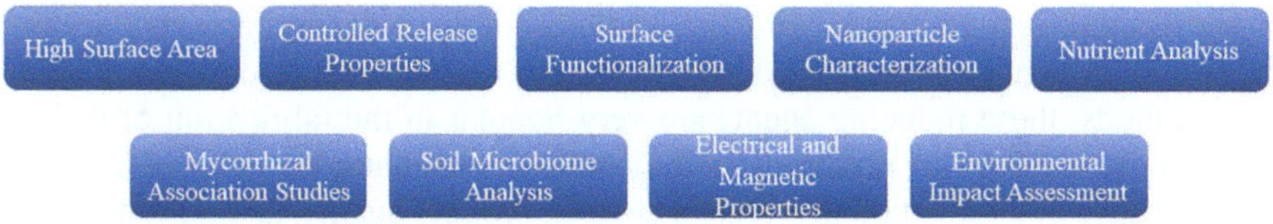

Fig. 1 Characteristics of Cu-based NMs

the mechanisms of action, merits and demerits, advantages, challenges, and future prospects of these Cu-based NMs in sustainable agriculture.

2 Chemical and Physical Characteristics of Cu-Based NMs

Cu-based NMs have broad surface area, catalytic activities, nutrient uptake, and delivery characteristics [14, 30, 62, 69]. CuNPs broad surface area develops the reactivity and catalytic characteristics in reactions like hydrogenation as well as oxidation [67]. Due to the inherent properties of Cu, these materials show mechanical strength, outstanding conductivity, and thermal stability [74]. The antimicrobial properties of Cu-based NMs make them very valuable in different biomedical applications [83]. By managing the size and shape of these nanoparticles, we can tailor the behavior of NPs, which influences the interactions of NPs with light, their interacting behavior with light is helpful in its utilization during the designing of sensors and other energy devices [21]. We are now using Cu-based NPs in techniques like soil integration, foliar spray, or hydroponic systems to deliver and uptake nutrients in soil and to plants. Various researchers are working actively and investigating the Cu-based NPs, their roles, and possible benefits and challenges during the delivery and uptake of nutrients in the soil as well as plants (Fig. 1).

3 Synthesis and Characterization of Cu-Based NMs

Various chemical reduction methods are used these days to synthesize CuNPs and CuONPs for the delivery and uptake of nutrients in soils. Methods like scanning electron microscopy, transmission electron microscopy, and X-ray diffraction are used to characterize the NPs for their size, shape, and chemical composition, while X-ray photoelectron spectroscopy (XPS) is used to analyze the surface properties of NPs. Using these methods, we can learn about the electrical and thermal conductivity, antibacterial activity, and catalytic efficiency of NMs. Information gained after extensive study about the applications of NMs range from sensors and remediation for environmental applications and electronics to energy storage devices [25]. Synthesis of Cu-based NMs encompasses methods like hydrothermal synthesis, chemical reduction, and green synthesis, giving various advantages in terms of additional material properties and scalabilities. Hydrazine, sodium borohydride, and ascorbic acid are the reducing agents which are used in the reduction of Cu salts by the chemical reduction methods, these reducing agents are very popular in the fabrication of CuNPs. These above methods are preferred for the synthesis and characterization of Cu-based NMs because of their simplicity, ease of control, product quality, and reproducibility [86]. For the synthesis of CuO and Cu (I) oxide (Cu_2O) nanostructures, generally hydrothermal and solvothermal methods are often used, which show greater photocatalytic and antibacterial activities [72]. Green synthesis routes of NPs are promoting these days, here plant extracts or, microorganisms have used for gaining eco-friendly alternatives. Using green synthesis methods, we can produce Cu-based NMs without toxic chemicals and having enhanced biocompatibility [54] (Table 1).

Characterization of Cu-based NMs provides valuable insights into the size distribution, crystallinity, and morphology of Cu nanoparticles [27]. The methods like XPS and Fourier-transform infrared spectroscopy (FTIR) are used to investigate the oxidation state, surface chemistry, and chemical bonding of Cu-based NMs [11]. The optical properties of Cu-based NMs evaluated by surface plasmon resonance

Table 1 Cu-based NMs and their description

Cu-based NMs	Description	Reference
Cu nanoparticles	Nanoscale particles of Cu with diameters greater than 100 nm	Flores et al. [20]
Cu nanowires	Long, thin structures with a diameter in the nanoscale typically have high aspect ratios	Guo et al. [24]
Cu nanotubes	Tubular structures composed of Cu at the nanoscale have varying diameters and lengths	Saifuddin et al. [53]
Cu nanorods	Rod-shaped nanoparticles with Cu as the primary component	Yang et al. [78]
Cu nanosheets	Thin, two-dimensional structures composed of Cu, often with nanoscale thickness	Xu et al. [75]
Cu oxide nanoparticles	Nanoparticles containing Cu and oxygen atoms, such as cuprous oxide (Cu_2O) and cupric oxide (CuO)	Momeni et al. [44]
Cu sulfide nanoparticles	NMs containing Cu and sulfur, such as Cu sulfide nanoparticles	Tetyana et al. [65]
Cu-based nanocomposites	Composite materials combine Cu with other NMs, polymers, or ceramics	Pozdnyakov et al. [49]
Cu quantum dots	Semiconductor nanocrystals with Cu atoms are often used in optoelectronic applications	Bhattacharyya et al. [3]
Cu nanocarriers	Nanoscale carriers or containers with Cu-based materials are used for drug delivery or imaging	He et al. [26]

Table 2 Different properties, description of Cu-based NMs

Property	Description	Examples of Cu-based NMs
Composition	Chemical makeup of the NMs	Cu nanoparticles, Cu oxide nanoparticles, and Cu-based nanocomposites
Particle size	Nanoscale dimensions for increased area and reactivity	Typically, less than 100 nm in diameter
Surface charge	Determines interactions with plant cells and nutrient ions	Can be engineered for positive or negative charges, influencing adsorption and uptake
Surface functionalization	Coating or modification to enhance stability and bioavailability	Organic ligands, polymers, or biomolecules can be used for functionalization
Solubility	Affects dispersibility in water or other nutrient solutions	Varies based on the specific nanomaterial and surface modifications
Stability	Resistance to agglomeration or degradation	Stable in various environmental conditions
Porosity	Influences nutrient loading and release	Porous structures allow for controlled release of nutrients
Magnetic properties	For targeted delivery or separation	Magnetic nanoparticles for controlled delivery
Photothermal properties	Ability to convert light into heat	Used for controlled release triggered by light
Biocompatibility	Compatibility with plant tissues	Biocompatible coatings to minimize toxicity
Nutrient adsorption capacity	Ability to adsorb and carry nutrients	Enhanced adsorption of specific ions or molecules
Controlled release capability	Ability to release nutrients in a controlled manner over time	Slow and sustained release for optimized uptake
Uptake efficiency	Efficiency of nutrient uptake by plants	Enhanced by NMs with favorable properties
Environmental impact	Considerations regarding potential toxicity and long-term effects on ecosystems	Ongoing research to understand and minimize any environmental impact

(SPR) and analyzed using UV-visible spectroscopy, and their catalytic efficiency can be evaluated and analyzed through techniques like cyclic voltammetry and gas chromatography technique [63]. Atomic force microscopy (AFM) and Brunauer–Emmett–Teller (BET) were used to determine the roughness and porosity of the NMs. Information about the roughness and porosity of the surface of NPs is essential for different applications in sensing and catalysis. So, in broad ways, Cu-based NMs offer a versatile and robust platform for a long range of applications and ongoing advancements in the synthesis and characterization of NMs, which are crucial while optimizing their properties for specific uses in our day-to-day lives. The continuous development in synthesis and characterization methods of Cu-based NMs will undoubtedly lead to the development of more efficient and sustainable Cu-based NMs in the future, which can resolve a range of issues related to nutrient delivery and uptake (Table 2).

4 Methods for Analyzing the Role of CuNMs in the Uptake and Delivery of Nutrients

Due to the high potential for delivering and uptake nutrients, Cu-based NMs improve agricultural productivity and sustainability. The use and investigation of Cu-based NMs for nutrient delivery and uptake in soil have been increasingly popular, and we hope they will likely establish any benchmark in the coming future. To assess the effectiveness and safety of these NMs, researchers are applying several approaches to use them, from laboratory-based assays to different field studies. The techniques like SEM, TEM, and XRD measurements are usually used for the characterization of Cu-based NMs to find information about their size, shape, and structural characteristics [38]. Features like size, shape, and structural characteristics develop the understanding of the mechanism that works under the interaction of the NMs with the soil particles as well as the root system of plants during the uptake and delivery of nutrients. Soil incubation experiments are the methods we are using these days for studying the behavior of Cu-based NMs in soils, these methods determine the bioavailability, mobility, and retention capacity of Cu-based NMs in the soil matrix. Some other methods, like inductively coupled plasma mass spectrometry (ICP-MS) or atomic absorption spectroscopy (AAS), are used these days for measuring the content of Cu in the soil and the plant tissues [59]. Monitoring the uptake and translocation of Cu into roots and whole plants and understanding and evaluating possible phytotoxicity would be very helpful in further studies. At the same time, we can start some pot and field experiments to observe the in-practice effects of Cu-based NMs on plant development and nutrient delivery and uptake. Using these experiments, formulations of the Cu-based NMs are directly applied to the soil, and then the efficiency of plants in the delivery and uptake of nutrients like phosphorus, nitrogen, and potassium. Some other methods, like fluorescence spectroscopy and XRF, are used to track the distribution of nutrients and supplement materials within the tissue of roots and plants [70]. Using 16S rRNA gene sequencing along with microbial respiration assays, researchers are confirming the introduction of Cu-based NMs disrupts or does not disrupt the health of soil and soil microbial community [31]. The modeling approaches like QSAR models as well as soil–plant transfer models are used for predictive drives on the way to the fate and transport of Cu-based NMs into the environment, which permits them to evaluate the long-term effect on nutrient cycling in addition to the yield of the crops [47]. Combined approaches help in thorough study

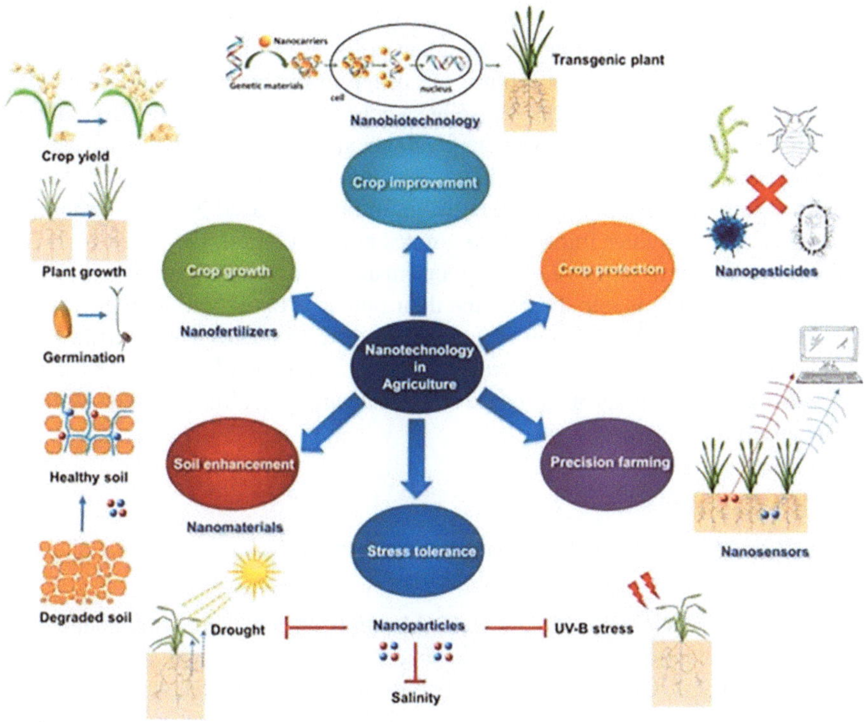

Fig. 2 Cu-based NMs used in nutrient delivery and uptake responsible to promote the growth, productivity of plants [55]

of Cu-based NMs about their potential influence on the progress of nutrient delivery to the soils and their environmental risks and benefits (Fig. 2).

4.1 Nutrient Formulation

Due to physicochemical properties, including high surface areas, excellent conductivity, and high antimicrobial activities Cu, Cu-based NMs are used for nutrient formulations. As per previous research Cu, Cu-based NMs promote higher bioavailability of nutrients, which are very important for plant growth while contributing to sustainable agricultural practices [15]. Better nutrient uptake and beneficial enzymatic activities in plants lead to increased crop yields, which can be facilitated by fertilizers enhancement through the incorporation of Cu-based NMs [28]. Furthermore, Cu-based NMs can be used as an effective delivery vehicle for nutritional supplements/micronutrients by providing a controlled release of nutrients and localized and targeted action, through these interventions, nutrient loss can be reduced, and plant productivity will be enhanced [87]. Apart from improving crop nutrition and growth, Cu-based NMs are also responsible for various benefits for agriculture and fighting soil-borne pathogens, bugs, and pests. By using antimicrobial properties of Cu-based NMs we can replace the usage of chemical pesticides by which eco-friendly agricultural practices have been achieved [68]. Additionally, Cu-based NMs are relatively nontoxic to the beneficial soil microorganisms when we use them at appropriate concentrations, which also supports their use in sustainable farming [60]. However, due to the highly reactive nature of Cu-based NMs, environmental conditions have a greater impact during nutrient formulation. Cu-based NMs offer significant benefits in terms of efficacy and sustainability, and their drawbacks, like the leaching of Cu ions into the soil and their potential toxicity to non-target organisms, can be solved using proper regulatory modification and management. Hence, recent research emphasizes optimizing the synthesis, characterization, uses, application, and disposal methods of Cu-based NMs to ensure safe, effective, and sustainable use in agriculture [50].

4.2 Plant Growth and Experimental Setup

In the last few years, the use of Cu-based NMs has been considered for enhancing the growth and production of plants by affecting several physiological and biochemical processes in plants. CuNPs have a highly enormous surface area, strong antimicrobial activity, and interaction capacity with different plant cells in several modes, which can be helpful in plant growth enhancement as well as the development of resistance in various stress conditions [51, 57]. Mishra et al. [43] and Siddiqui et al. [61] reported that NMs could be synthesized using various methods, including chemical reduction and laser ablation as well as their green synthesis using plant extracts and their further modifications to reduce their toxicity as well as compatibility in a plant system. CuNPs-mediated plant growth can be examined, where nanoparticle suspensions in diverse concentrations are directly applied to the plants by spraying leaves, soil amended, and seed treatment [66]. The impact of CuNPs on plant growth can be measured and accessed by seed germination, root and shoot length, chlorophyll content, and biomass production [48]. Nair et al. [45] and Sharma et al. [58] reported that CuNPs can enhance photosynthetic activity, simplify nutrient delivery and uptake, and upsurge antioxidant activity in plants during normal as well as in stress. Ghorbanpour et al. [22] reported that the concentration of CuNPs is very significant, high concentrations may lead to phytotoxicity, stunted growth of the plants, as well as cause chlorosis in plants. CuNPs can penetrate root cell walls and moderate plant development by triggering certain signaling pathways, inducing related expression of growth gene and improving tolerance against the stress [32]. Some reports support the potential of CuNPs in plant resistance to environmental stresses like drought, salinity, and heavy metal toxicity [33, 40]. Mechanisms that underpin the effects of CuNPs on plant systems are still under investigation, and the impact of long-term implications of Cu-based NMs on plant health and soil ecology remains under examination.

4.3 Nutrient Delivery Experiments

CuNPs are also useful in hydroponic systems for nutrient uptake and delivery to plants. Nutrient delivery focuses on optimizing fertilizer application, nutrient use efficiency, and reducing adverse environmental impacts from nutrient runoff [81]. Nutrient delivery examination in animals investigates the impact of various diets and nutrient timing on nutrient absorption and metabolism [41]. Microbial nutrient delivery experiments involve soil microorganisms to enhance nutrient cycling and upsurge soil health for sustainable agriculture [7]. NPs are also implemented to enhance smart delivery systems where nano-carriers are used to encapsulate nutrients to ensure localized and controlled release, improving bioavailability and decreasing losses [80]. For the improvement of agricultural practices, human health, and environmental sustainability, the implementation of nutrient delivery experiments is very crucial.

4.4 Nutrient Uptake Analysis

Inductively Coupled Plasma-Mass Spectrometry (ICP-MS) techniques are very useful in determining the nutrient delivery absorption and concentrations of nutrients within the plant tissues. The uptake of nutrients in the presence of Cu-based NMs has increasingly received attention due to their vast applications in agricultural, environmental remediation, and biomedical sectors. As per the previously published report, CuNPs stimulate the uptake of nutrients like nitrogen, phosphorus, and potassium, which are essential for crop growth and production [18]. However, an overabundance of Cu accumulation may also lead to tissue toxicity, resulting in nutrient uptake dysfunction [23]. Cu-based NMs also facilitate the nutrient supply across the cell membrane, which holds the potential for improving the health of the plants in nutrient-deficient conditions [82]. The interaction of soil microorganisms with CuNPs plays an important role in inducing nutrient cycling and their availability in the plant rhizosphere [58]. The actual effects of Cu-based NMs on nutrient uptake and delivery will be better understood in relation to optimizing their utilization and avoiding unintended ecological consequences.

4.5 Data Analysis

Synthesis, characterization, and analysis of Cu-based NMs depends on various methods, including chemical reduction, laser ablation, and sol–gel techniques, to produce controlled size and shape directly affecting performance in various applications [37, 80]. For studying the structural properties, morphology and crystallinity of Cu-based NMs, techniques like TEM, SEM, and XRD are the mainstay and their roles are very vital [13] as well as determining the chemical states and optical properties role of XPS and UV-spectroscopy are very crucial [84]. Data integrates Cu-based NMs' properties at the structural, optical, and electrochemical levels to measure the performance of materials under long range of applications. Further analysis of these datasets in Cu-based NMs research identifies trends and associations between circumstances during production and resultant attributes of material systems that lead to improved design for forthcoming technologies [72].

4.6 Safety and Environmental Considerations

Due to high surface area and sensitivity, CuNPs may cause various risks related to human health, like inhalation toxicity, adsorption through the skin, lung injury, and liver damage [46]. Wearing the appropriate PPE should be necessary during the processes' synthesis, handling, and sandsal. These NMs can leach into the environment, disrupting ecological balance and causing toxicity to aquatic organisms [64]. To reduce these risks, green synthesis methods are essential because only environmentally friendly solvents are used during green synthesis, reducing the formation of hazardous by-products [34]. Sound safety procedures, combined with environmentally conscious formation and proper disposal of NMs, should assure that Cu-based NMs' advantages are not counterpoised by human well-being or environmental degradation.

5 Uses of Cu-Based NMs in Delivery and Uptake of Nutrients

Cu-based NPs have a range of applications in delivering as well as absorbing nutrients, which significantly put their impact on agricultural and environmental systems [73]. Due to unique physicochemical properties like high surface area, biocompatibility, and tunable reactivity, Cu-based NMs are very promising in nutrient uptake and delivery applications. The engineered CuNPs, CuO NPs, and Cu-based nanocomposites enhance the bioavailability of nutrients as well as also increase the therapeutic efficacy by controlled release of nutrients. The CuNPs have some advantageous characteristics that improve the solubility and stability of the nutrients and are hydrophobic in nature, like vitamins, antioxidants, and essential minerals [10, 12]. Cu-based NMs can also help enhance cellular uptake by exploiting their interaction with the cell membrane. Different ligands or polymers have been explored for the surface modification of CuNPs to enhance the interaction of specific receptors or stability in biological environments [71]. CuNPs also enhance the bioavailability of micronutrients to the plants and support nutrient uptake from the soil to crop for production improvement [35]. Additionally, Cu-based NMs may provide additional

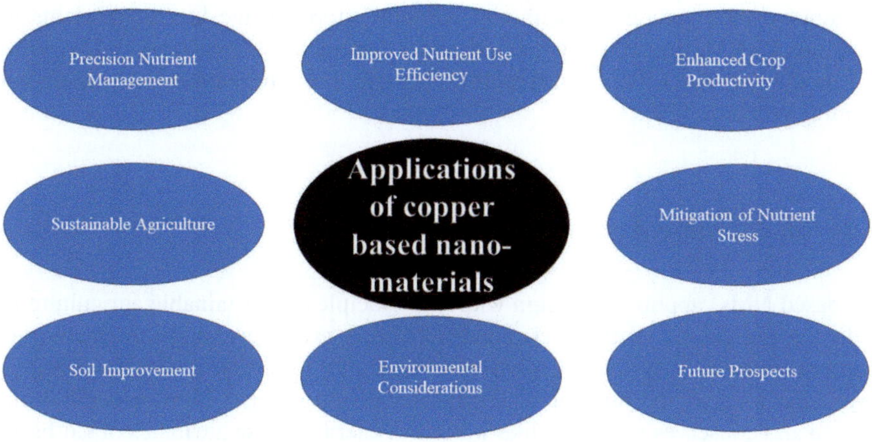

Fig. 3 Applications of Cu-based NMs

health benefits by developing products of dietary supplements or functional food with their antimicrobial, antioxidant, and anti-inflammatory properties. Zhang et al. [80] reported that CuO NPs have the potential to increase health benefits through nutrient administration, overcome oxidative stress (ROS) as well as supporting immunological responses. Cu-based NMs provide the best balance of nutrients to the plants for their growth and also optimize the delivery of nutrients in hydroponic systems and controlled agricultural environments (Fig. 3).

5.1 Nutrient Management

Cu-based NMs deliver a localized and accurate approach to nutrient management that can address the inefficiencies of conventional fertilizers. The nanoscale properties of CuNMS facilitate the controlled release and delivery of nutrients to ensure the optimal supply of nutrients over time. By reducing nutrient wastage, reducing runoff minerals into water bodies, and enhancing the yield of agricultural crops, CuNMS can also deliver micronutrients like copper, which can give additional support to plant metabolic activities and development [77].

5.2 Nutrient Use Efficiency

CuNPs significantly increase the efficiency of using the nutrients by improving their absorption, retention as well as utilization capacity of nutrients. The high surface area-to-volume ratio of the Cu-based NMs and their surface functionalization help to improve their interaction with soil particles, stopping the loss of nutrients by leaching. CuNMs are reducing the dependency on chemical fertilizers and lowering the production costs that influence cost-effective and environmentally sustainable farming practices [19].

5.3 Sustainable Agriculture

Cu-based NMs' applications align with the principles of sustainable agriculture by reducing the use of synthetic/chemical fertilizers and decreasing the leaching of nutrients into water systems. Cu-based NMs' ability to release controlled nutrients over time reduces the number of fertilizer applications in a farm throughout the season and conserves resources like water and energy. These attributes of Cu-based NMs support environmentally friendly agriculture practices while maintaining a high agricultural yield [42].

5.4 Effects on Plant Growth

CuNMs may stimulate seed germination as well as also enhance the early growth stages. CuNPs also facilitate the uptake of nutrients, encouraging root and shoot development in plants [85]. Cu is a cofactor for various photosynthesis enzymes like cytochrome c oxidase, plastocyanin. CuNMs may also improve the bioavailability of Cu in the plant, which can increase the photosynthetic efficacy, resulting in complete growth of the plants [36].

5.5 Impact on Plant Yield

Cu-based NM application in optimal concentrations has raised the production of some crops by increasing their nutrient uptake and encouraging plant growth. CuNPs may boost rice production, and as per Zhao et al. [83–85], CuNMs may increase the tomato fruit size. CuO NPs have antimicrobial properties that can protect plants from certain diseases caused by microbes and improve crop production indirectly. CuO NPs also work as preventive agents against various fungal diseases of tomatoes and cucumbers [1].

6 Conclusions

Cu-based NMs play a vital role in the uptake and delivery of nutrients to the soil and roots of the plant for better growth and improved production. The biocompatibility, high surface area, and ease of functionalization make Cu-based NMs an ideal contender for targeted and efficient nutrient transport systems. CuNPs can improve plants' absorption of essential nutrients, which leads to better growth and production and alleviates resistance to stress and infections. As continuous advancements in nanotechnology, optimizing and boosting the use of Cu-based NMs in nutrient uptake and delivery will likely play a significant and vital role in addressing global challenges related to food security and sustainable agriculture.

Competing Interests The authors declare that they have no competing interests.

Declaration of Generative AI and AI-Assisted Technologies in the Writing Process During the preparation of this work the book chapter author(s) have not taken any assistance from AI, and the book chapter author(s) reviewed and edited the content as needed and take(s) full responsibility for the content of the publication. All authors read and approved the final version of the manuscript.

References

1. Amin MA, Ameen M, Lee JH (2020) Copper-based nanomaterials in agriculture: benefits and potential risks. Environ Toxicol Pharmacol 76:103363
2. An C, Sun C, Li N (2022) NMs and nanotechnology for the delivery of agrochemicals: strategies towards sustainable agriculture. J Nanobiotechnol 20(1):11
3. Bhattacharyya B, Balischewski C, Pacholski C, Pandey A, Bald I, Taubert A (2023) Cu iron chalcogenide semiconductor nanocrystals in energy and optoelectronics applications state of the art, challenges, and future potential. Adv Optic Mater 11:8
4. Bachheti RK, Bachheti A, Husen A (eds) (2023) Nanomaterials for environmental and agricultural sectors. Springer, Singapore. https://doi.org/10.1007/978-981-99-2874-3
5. Bachheti A, Bachheti RK, Husen A (eds) (2024) Carbon-based nanomaterials: synthesis, agricultural, biomedical, and environmental interventions. Springer, Singapore. https://doi.org/10.1007/978-981-97-0240-4
6. Bachheti RK, Bachheti A, Husen A (eds) (2024a) Metal and metal-oxide based nanomaterials. In: Synthesis, agricultural, biomedical and environmental interventions. Springer, Singapore. https://doi.org/10.1007/978-981-99-7673-7
7. Berendsen RL, Pieterse CMJ, Bakker PAHM (2012) The rhizosphere microbiome and plant health. Trends Plant Sci 17(8):478–486. https://doi.org/10.1016/j.tplants.2012.04.001
8. Bharathi S, Suman S, Durga M (2017) Cu-based nanoparticles for plant nutrient delivery: Current trends and future prospects. Environ Sci Pollut Res 24(15):13955–13964. https://doi.org/10.1007/s11356-017-9396-6
9. Cakmak I (2000) Role of zinc in protecting plant cells from reactive oxygen species. New Phytol 146(2):185–205. https://doi.org/10.1046/j.1469-8137.2000.00630.x
10. Cai X, Zhang H, Liu Y, Guo Y (2021) Cu-based nanoparticles for controlled release of bioactive compounds and their application in nutrient delivery. J Nanobiotechnol 19(1):1–14
11. Chen W et al (2021) Characterization of Cu-based NMs for catalytic applications. J Nanosci Nanotechnol 21(2):45–52

12. Chen Z, Wang L, Li J, Zhang Y (2022) Development of Cu nanoparticles for enhanced bioavailability and stability of vitamins in food formulations. Food Sci Techno 53(2):1357–1365
13. Cheng Y, Zhang Z, Lee H (2018) Synthesis and characterization of Cu nanoparticles for catalysis and sensing applications. J Nanosci Nanotechnol 18(4):2453–2461
14. Crisan MC, Mocan T, Lucian M (2021) Cu nanoparticles: synthesis and characterization, physiology, toxicity and antimicrobial applications. Appl Sci 12(1):141
15. Cui Y, Wang Z, Liu H (2019) Cu-based NMs in plant growth promotion and stress mitigation: current applications and future prospects. J Agric Food Chem 67(18):5068–5077
16. Dinesh R, Chandra R, Raja M (2020) Cu oxide nanoparticles as a carrier for micronutrient delivery to plants: a review. J Nanosci Nanotechnol 20(8):5045–5055. https://doi.org/10.1166/jnn.2020.17617
17. Dobermann A, Bruulsema T, Çakmak I, Gerard B, Majumdar K, McLaughlin MJ, Zhang X (2022) Responsible plant nutrition: a new paradigm to support food system transformation. Glob Food Secur 33:100636
18. El-Temsah YS, Joner EJ (2015) The effect of Cu nanoparticles on the nutrient uptake and growth of plants. Environ Toxicol Chem 34(8):1920–1927
19. El-Saadony MT, ALmoshadak AS, Shafi ME, Albaqami NM, Saad AM, El-Tahan AM, Desoky ES, Elnahal AS, Almakas A, Abd El-Mageed TA, Taha AE (2021) Vital roles of sustainable nano-fertilizers in improving plant quality and quantity-an updated review. Saudi J Biol Sci 28(12):7349–7359
20. Flores LH, Debut A, Stael C (2015) Synthesis of multicomponent nanoparticles for immobi lization of heavy metals in aqueous phase. NanoWorld J 1(2)
21. Gao L et al (2017) Optical and electrical properties of Cu nanoparticles for advanced device applications. Nanotechnol 28(9):095702
22. Ghorbanpour M, Sadegh HR, Farahani HA (2016) Cu nanoparticles (CuNPs) as a growth enhancer for plants: a review. Environ Sci Pollut Res 23(2):1392–1406
23. Gomes ME et al (2020) Impact of Cu nanoparticles on plant growth, nutrient uptake, and soil microbial activity. Environ Sci Pollut Res 27(30):37228–37239
24. Guo H, Lin N, Chen Y (2013) Cu nanowires as fully transparent conductive electrodes. Sci Rep 3:2323
25. Gupta V et al (2020) Cu NMs: synthesis, properties, and applications. Mater Sci Eng R Rep 142:1–14
26. He S, Wu L, Li X, Sun H, Xiong T, Liu J, Zhang J (2021) Metal-organic frameworks for advanced drug delivery. Acta Pharmaceutica Sinica B 11(8):2362–2395
27. Huang M et al (2019) Surface characterization of Cu nanoparticles for improved antibacterial activity. J Mater Chem B 7(3):251–260
28. Huang Q, Zhang X, Xu Y (2020) Role of Cu nanoparticles in enhancing the growth and yield of crops. Nanomater 10(3):492
29. Azamal H, Bachheti RK, Bachheti A (eds) (2023) Current trends in green nano-emulsions: food, agriculture and biomedical sectors. Springer, Singapore. https://doi.org/10.1007/978-981-99-5398-1
30. Joshi A, Sharma A, Bachheti RK, Husen A, Mishra VK (2019) Plant-mediated synthesis of copper oxide nanoparticles and their biological applications. In: Nanomaterials and plant potential. Springer, Cham, pp 221–237. https://doi.org/10.1007/978-3-030-05569-1_8
31. Kaiser J, Zhang Z, Li S (2022) The effect of Cu-based nanoparticles on soil microbial diversity and activity. Environ Sci Pollut Res 29(15):22001–22015
32. Khalil SA, Abbasi BH, Shah Z (2017) Effects of Cu nanoparticles on plant growth and development: a review. Environ Toxicol Pharmacol 51:18–28
33. Khan Z, Haider F, Ali B (2021) Cu nanoparticles in plants: a comprehensive review on effects and mechanisms. Front Plant Sci 12:660893
34. Keller AA, Wang H, Zhao L (2013) Environmental release, fate, and effects of engineered NMs: a critical review. Environ Toxicol Chem 32(7):1459–1474. https://doi.org/10.1002/etc.2231

35. Kumar P, Singh N, Thakur S (2019) Cu nanoparticles in agriculture: Implications for nutrient uptake and plant growth promotion. J Agric Food Chem 67(4):927–936
36. Kumar S, Kumar R, Singh P (2020) Copper nanoparticles: a potent tool for enhancing agricultural productivity and plant health. Sci Total Environ 729:138704
37. Liu J, Li X, Yang Y (2017) Cu NMs: synthesis, properties, and applications in catalysis. J Mater Sci 52(11):6467–6479
38. Liu X, Wu F, Chen Z (2020) Synthesis and characterization of Cu-based NMs for agricultural applications. J Nanosci Nanotechnol 20(4):2345–2352
39. Ma X, Geiser-Lee J, Deng Y, Kolmakov A (2017) Interactions between engineered nanoparticles (ENPs) and plants: phytotoxicity, uptake, and accumulation. Sci Total Environ 407(18):4975–4982
40. Mahmoudi N, Khalil SA, Dighriri AA (2017) Cu nanoparticles in agriculture: a future avenue for enhancing plant growth. Plant Growth Regul 83(3):527–541
41. Miller GL, McDonald AC, Thomas ML (2018) Nutrient timing and the gut microbiota: implications for human health. Nutr Res Rev 31(1):22–34. https://doi.org/10.1017/S0954422417000202
42. Mishra P, Singh A, Mishra R, Ong S, Prasad SM (2018) Application of nanotechnology to enhance the nutrient quality of food crops and agricultural production. In: Nanomaterials in plants algae and microorganisms, pp 453–472
43. Mishra S et al (2020) Cu nanoparticles as nanocarriers for enhanced nutrient delivery in plants. J Nanobiotechnol 18(1):1–11. https://doi.org/10.1186/s12951-020-00697-9
44. Momeni S, Sedaghati F (2018) Cuo/cu2o nanoparticles: a simple and green synthesis, characterization and their electrocatalytic performance toward formaldehyde oxidation. Microchem J 143:64–71
45. Nair R, Varghese SH, Nair BG (2014) Cu-based nanoparticles for improving plant growth and stress tolerance. Environ Nanotechnol Monit 1:34–41
46. Nel A, Madler L, Velegol D (2006) Toxicology and environmental impact of NMs. Nanotechnol 17(6):2157–2172. https://doi.org/10.1088/0957-4484/17/6/007
47. Pallavi S, Gupta P, Yadav A (2023) Modeling the impact of Cu nanoparticles on nutrient dynamics in soils. Environ Model Assess 28(3):245–258
48. Perelshtein I, Perkas N, Gedanken A (2016) Cu nanoparticles as potential agents for plant growth enhancement. Environ Sci Technol 50(18):9847–9856
49. Pozdnyakov AS, Emelyanov AI, Korzhova SA, Kuznetsova NP, Bolgova YI, Trofimova OM, Semenova TA, Prozorova GF (2021) Green synthesis of stable nanocomposites containing Cu nanoparticles incorporated in Poly-N-vinylimidazole. Polymers 13:3212
50. Rai M, Yadav A, Gade A (2021) Cu-based NMs in agriculture: the impact of their application on plant health and productivity. Nanosci Nanotechnol Lett 13(9):687–695
51. Raliya R, Tarafdar JC, Biswas P (2015) Role of Cu nanoparticles in plant growth and development. Plant Stress 8(1):1–8
52. Rizwan M, Ali S, Adrees M, Rizvi H, Zia-ur-Rehman M, Hannan F (2019) Cadmium stress in rice: toxic effects, tolerance mechanisms, and management: a critical review. Environ Sci Pollut Res 26(15):14620–14633
53. Saifuddin N, Raziah AZ, Junizah AR (2018) Carbon Nanotubes: A Review on Structure and Their Interaction with Proteins. J Chem 18:676815
54. Sathishkumar M et al (2016) Green synthesis of Cu nanoparticles: a review. J Nanosci Nanotechnol 16(6):760–773
55. Shang Y, Hasan MK, Ahammed GJ, Li M, Yin H, Zhou J (2019) Applications of nanotechnology in plant growth and crop protection: a review. Molecules 24(14):2558. https://doi.org/10.3390/molecules24142558
56. Sharma VK, Yngard RA, Lin Y (2012) Cu nanoparticles: synthesis, properties, and applications. J Phys Chem C 116(10):5063–5073. https://doi.org/10.1021/jp211923v
57. Sharma V, Patel M, Chattopadhyay S (2017a) Effect of Cu nanoparticles on seed germination and plant growth: a review. Environ Toxicol Pharmacol 52:135–144

58. Sharma VK et al (2017b) NMs for environmental remediation and nutrient uptake in plants: a review. Environ Nanotechnol Mon Manage 8:12–21
59. Sharma V, Yadav R, Meena R (2021) Cu NMs in agriculture: Fate and transport in soils. Sci Total Environ 769:144302
60. Sharma S, Suresh M, Gupta R (2022) Cu nanoparticles in agricultural applications: from synthesis to environmental interactions. Sci Total Environ 807:150710
61. Siddiqui MH, Al-Whaibi MH, Basalah MO (2020) Cu nanoparticles and their role in plant growth promotion: a review. Environ Sci Pollut Res 27(9):9481–9494
62. Siddiqi KS, Husen A (2020) Current status of plant metabolite-based fabrication of copper/copper oxide nanoparticles and their applications: a review. Biomater Res 24(11):1–15. https://doi.org/10.1186/s40824-020-00188-1
63. Singh R et al (2019) Electrochemical and photocatalytic properties of Cu-based NMs. J Appl Electrochem 49(1):15–27
64. Sundrarajan M, Sivasubramanian V, Shanmugam S (2015) Cu nanoparticle-based antimicrobial materials and applications. J Mater Sci 50(2):101–113. https://doi.org/10.1007/s10853-014-8609-2
65. Tetyana P, Mphuthi N, Jijana AN, Moloto N, Shumbula PM, Skepu A, Vilakazi LS, Sikhwivhilu L (2021) Synthesis, characterization, and electrochemical evaluation of Cu sulfide nanoparticles and their application for non-enzymatic glucose detection in blood samples. Nanomater 13:481
66. Tripathi DK, Srivastava PK, Singh S (2017) Impact of Cu nanoparticles on plant growth and development. Ecotoxicol 26(8):711–723
67. Vasquez MI et al (2015) Cu nanoparticles as catalysts for organic reactions: mechanisms and applications. J Nanosci Nanotechnol 15(7):5280–5295
68. Vandebriel RJ, de Jong WH (2020) Nano-silver and Cu-based materials: a systematic review of their toxicity to plants and microorganisms. Ecotoxicol 29(3):303–321
69. Wassie AT, Ahmed IN, Bachheti A, Husen A, Bachheti RK (2024) Plant extract-assisted green synthesis and characterization of copper nanocomposite. In: Metal and metal-oxide based nanomaterials: synthesis, agricultural, biomedical and environmental interventions. Springer, Singapore, pp 43–66. https://doi.org/10.1007/978-981-99-7673-7_3
70. Wang J, Zhang P, Li L (2019) Tracking nutrient uptake in plants treated with Cu nanoparticles using X-ray fluorescence and fluorescence spectroscopy. Plant Soil 439(1–2):143–155
71. Wang W, Li J, Zhang X (2020a) The environmental impact of Cu nanoparticles in soil and plants: mechanisms and implications. Environ Pollut 265:114784. https://doi.org/10.1016/j.envpol.2020.114784
72. Wang X et al (2021a) Hydrothermal synthesis of Cu oxide nanostructures and their photocatalytic applications. J Nanotechnol 32(4):315–324
73. Woźniak-Budych MJ, Staszak K, Staszak M (2023) Cu and Cu-based nanoparticles in medicine-perspectives and challenges. Molecules (Basel, Switzerland) 28(18):6687
74. Wu J et al (2016) Cu-based NMs for thermal and electrical applications. Mater Sci Eng R 107:1–11
75. Xu Y, Chen D, Jiao X, Xue K (2007) Cuo microflowers composed of nanosheets: synthesis, characterization, and formation mechanism. Mater Res Bull 42(9):1723–1731
76. Yadav RK, Parihar P, Yadav G, Tiwari R, Parihar L, Lary DP (2020) Nano-fertilization for sustainable crop production: a review. In: NMs in plants algae and microorganisms, pp 285–305
77. Yadav A, Yadav K, Abd-Elsalam KA (2023) Nanofertilizers: types, delivery and advantages in agricultural sustainability. Agrochem 2:296–336
78. Yang X, Chen S, Zhao S, Li D, Ma H (2003) Synthesis of Cu nanorods using electrochemical methods. J Serbian Chem Soc 68(11):843–847
79. Yang Y, Zhang T, Zhang S (2013) Cu-based nanoparticles for plant growth enhancement and disease resistance: mechanisms and applications. Environ Toxicol Chem 32(9):2106–2116. https://doi.org/10.1002/etc.2305
80. Zhang J, Lee S, Gao L (2020b) Cu-based NMs: fabrication, properties, and applications. Mater Sci Eng R 136:52–73

81. Zhao Y, Zhang F, Sun Y (2016a) Optimizing nutrient delivery in agricultural systems for improved environmental sustainability. Environ Sci Technol 50(6):3013–3020. https://doi.org/10.1021/acs.est.5b04934
82. Zhao L et al (2016) Cu-based NMs: their interactions with plants and potential for agricultural applications. Sci Total Environ 550:928–939
83. Zhao X et al (2019a) Antibacterial properties of Cu nanoparticles: mechanisms and applications. Front Microbiol 10:788
84. Zhao X, Li P, Wang Z (2019b) Optical and electrochemical properties of Cu nanoparticles: analysis and application. Nano Let 19(3):1450–1457
85. Zhao Y, Li Y, Han S (2019c) Effects of copper nanoparticles on plant growth and development: a review. Sci Total Environ 666:1062–1073. https://doi.org/10.1016/j.scitotenv.2019.02.193
86. Zhao X et al (2020) Synthesis of cu nanoparticles and their antimicrobial properties. Mater Sci Eng C 116:111105
87. Zhou M, Liu Q, Zhang Z (2021) Cu nanoparticles in sustainable agriculture: a review of their potential applications and environmental risks. Environ Toxicol Chem 40(4):1015–1026

Use of Copper-Based Nanomaterials in Seed Germination and Plant Growth

Deepanjan Mridha, Archita Dey, Pascal Labrousse, and Tarit Roychowdhury

Abstract Nanotechnology is an emerging field dedicated to manipulating atoms and molecules and demonstrates significant potential across various scientific domains. This innovative discipline focuses on nanoparticles (NPs) with diameters ranging from 1 to 100 nm, capitalizing on their small size and extensive surface area to enhance reactivity. Out of which, copper-based nanomaterial is one of the most used in nanotechnology. Copper and copper-conjugated nanoparticles, synthesized through green approaches, offer economic and eco-conscious advantages. These copper nanoparticles find diverse applications in biomedicine, pharmaceuticals, bioremediation, molecular biology, bioengineering, genetic engineering, dye degradation, catalysis, cosmetics, and textiles. This chapter overviews on the effects and applications of copper nanoparticles, primarily emphasizing green synthesis of copper nanoparticles following their application in seed germination and plant growth.

Keywords Nanotechnology · Copper nanoparticle · Green synthesis · Seed germination · Plant growth

D. Mridha (✉)
Department of Crop and Soil Sciences, Washington State University, Pullman, WA, USA
e-mail: mridha.deepanjan@wsu.edu; mridha.deepanjan@gmail.com

D. Mridha · A. Dey · T. Roychowdhury
School of Environmental Studies, Jadavpur University, Kolkata, India

P. Labrousse
Université de Limoges, E2Lim UR 24133, Limoges, France

R. K. Bachheti et al. (eds.), *Copper-Based Nanomaterial for Agricultural Practices*, Smart Nanomaterials Technology, https://doi.org/10.1007/978-981-96-8610-0_8

1 Introduction

In the past few years, there has been a significant increase in interest regarding the application of nanotechnology for creating materials and delivery systems at the nanoscale, aiming to tackle the gradual decline in the effectiveness of mineral fertilization (Lowry et al. 2019). Nanotechnology utilizes nanoparticles (NPs), particularly engineered nanoparticles (ENPs), crafted through synthesis and manipulation at atomic, molecular, and macromolecular levels [1]. Nanomaterials (NMs) are defined by having at least one dimension within the nanometer (nm) range. A nanometer is equivalent to one billionth of a meter (10–9 m), representing a length comparable to 10 hydrogen or 5 silicon atoms arranged in a line (Husen and Iqbal 2019). The utilization of nanoparticle-based (NP-based) systems for delivering pesticides and fertilizers can potentially diminish the ecological impact of agricultural practices. Additionally, NP-based methods for delivering genetic materials and nutraceuticals may enhance crop yield and fortify plant resilience to diverse environmental stressors, including plant pathogens. Lastly, incorporating these nanoparticles in soil and water remediation and conservation efforts could restore these dwindling resources, ultimately benefiting agricultural production (Bakshi and Abhilash 2020). Nevertheless, these nanomaterials exhibit greater reactivity and bioavailability due to their inherent characteristics than conventional forms commonly employed in agriculture (Adisa et al. 2019). The term "nanomaterial" (NM) denotes a material with at least one dimension measuring less than 100 nm (Szőllősi et al. 2020). With the advancement of nanotechnology, there is a notable surge in the utilization of nanomaterials. Research efforts are imperative to examine the impacts of nanomaterials on all living organisms, particularly sessile plants that cannot evade these external factors. Given their escalating emissions, it is crucial to assess the potential hazards of nanomaterials in the environment and their effects on plants, animals, and humans. Among the reported nanomaterials to date, copper metal and nanomaterials stand out due to their cost-effectiveness, abundance on the Earth, and versatile properties, including thermal and heat conductivity, ductility, strength, magnetism, antimicrobial activity, and electrical conductivity [2]. Optimal copper (Cu) nanomaterials can be achieved by managing the size of bulk materials within the range of 1–100 nm while preserving essential properties like size, stability, surface area, shape, and crystal structure. Copper and copper-based nanomaterials exhibit unique properties compared to their bulk counterparts, owing to their heightened surface area and distinctive quantum characteristics (Gawande et al. 2016; [3, 4]). Copper and copper-based nanomaterials are particularly appealing due to their abundant availability and cost-effectiveness compared to noble metals. Moreover, these nanomaterials demonstrate outstanding efficacy in various applications, including electro-catalysis, solar/photovoltaic energy conversion, photochemical catalysis, and catalytic applications within the realms of industries and pharmaceuticals (Tonelli and Gualandi 2019). The synthesis of nanomaterials can be categorized into two primary groups: traditional methods and green methods. Traditional nanomaterial synthesis methods present various advantages, including the production of a wide range of nanoparticles suitable for numerous applications. Certain methods offer significant scalability [5], precise control over nanoparticle morphology [6], and applications in innovative fields such as battery conduction, electrical applications [7], targeted disease therapy [8], and energy storage/conservation [9]. Nevertheless, the significant adverse consequences of applying these traditional methods cannot be ignored. The widespread use of organic solvents in nanomaterial synthesis presents a substantial risk to neurobehavioral and reproductive health during synthesis [10]. Furthermore, the utilization of high pressure and elevated temperatures may contribute to hazardous working conditions [11]. Another crucial concern arises from the release of volatile vapors [12] and the excessive production of carbon dioxide, which significantly contributes to the greenhouse effect, representing a top priority adverse effect stemming from these synthesis processes [13]. These approaches present inherent risks to both the scientists involved in the synthesis process and the environment. The potential hazards associated with these methods surpass the advantages offered by traditional nanomaterial synthesis. As a result of these considerations, traditional synthesis approaches have lost favor, giving rise to the emergence of green synthesis. Given the ongoing climate crisis, developing novel and forward-thinking methods aligned with the 12 Principles of Green Chemistry is crucial. Green synthesis utilizes a clean, safe, cost-effective, and environmentally friendly approach to fabricate nanomaterials. Microorganisms, including bacteria, yeast, fungi, algal species, and specific plants, serve as substrates for the green synthesis of nanomaterials. Various active molecules and precursors, such as metal salts influence the ultimate morphology and size of the nanoparticles. Furthermore, green synthesis imparts nanomaterial benefits that extend from antimicrobial properties [14] to natural reducing and stabilizing properties. The active molecules found in microorganisms used as substrates for green synthesis contribute to these properties—a recent revelation since the last comprehensive nanomaterial review published by Saratale et al. [15]. The components of the green species employed in nanomaterial synthesis often include specific enzymes [16], amino acid groups [17], proteins [18], or distinct chemical structures. The green approach has found extensive application in synthesizing copper nanoparticles (CuNPs) due to its cost-effectiveness, cleanliness, non-toxicity, and absence of pollution (Roy et al. 2019). This method follows a bottom-up approach for CuNPs preparation, incorporating a subsequent redox reaction [19]. A green source, such as a plant extract, serves as both the reducing agent and stabilizer in CuNPs preparation, preventing aggregation and mitigating the formation of potentially harmful by-products (128). This approach facilitates environmentally friendly reaction conditions for the production of CuNPs.

Crucially, copper plays a crucial role as a vital micronutrient, influencing numerous essential physiological processes in plants. These processes encompass enzymatic activity, photosynthesis, and cellular functions, including cell wall formation, oxidative stress response, mitochondrial respiration, and protein metabolism. Copper nanoparticles (CuNPs) are some of the most significant and extensively manufactured metal nanoparticles. CuNPs are advantageous to other more stable metal nanoparticles for their superior conductivity (Bakshi and Kumar 2021). Copper, as an element, becomes toxic beyond a certain threshold level, and this threshold varies depending on the type of crop plants [20]. Copper nanoparticles manifest as brownish-black powders and undergo reduction upon exposure to carbon monoxide or hydrogen, harming humans and threatening aquatic life (Rajput et al. 2018). Several recent studies have showcased the distinctive capabilities of copper-based nanomaterials in agriculture.

Field investigations have illustrated the effectiveness of materials like $Cu_3(PO_4)_2$-based nanosheets and commercial CuO nanoparticles when applied to young seedlings through foliar application. These applications have demonstrated enhanced plant growth and the suppression of Fusarium spp. infections in tomatoes (*Solanum lycopersicum* L.) and watermelons (*Citrullus lanatus* L.) throughout their entire life cycles (Mamiroli et al. 2021). Negative impacts on crop plants have also been documented, such as biochemical changes, oxidative stress, and reduction of root and shoot development and seed germination. These differences may result from particles' capacity to pass through plant cell membranes. The ambiguity around how the characteristics of nanoparticles change after they interact with the environment and how environmental factors impact nanomaterials presents one of the major obstacles to determining the risk associated with a nanomaterial release event. One of the most challenging issues in defining risk associated with a nanomaterial release event is the ambiguity about how the characteristics of nanomaterials change after they interact with the environment and how meteorological conditions impact nanomaterials. Thus, it is necessary to evaluate the threats that NPs pose to the environment. Metal and metal oxide NPs, in particular Cu-based nanomaterials, are receiving special attention due to the growing interest in the effect of NPs. This chapter provides an in-depth review of the synthesis of Cu-based NPs, emphasizing green/biosynthetic NPs, their impact on agricultural plant germination and growth, and how they alleviate plant stresses.

2 Copper Nanoparticles (CuNPs) and Its Synthesis

Plant tissues contain a large amount of copper, an essential micronutrient necessary for growth and involved in several physiological and metabolic processes [21]. It is a component of regulatory proteins, a cofactor of ascorbate oxidase and superoxide phenol oxidases, and it participates in the electron transport chain (ETC) during photosynthesis and respiration [22]. It has been found that around 10 mg of Cu are present in 1 kg of dry plant tissue on average [23]. Its application in many sectors has, on the one hand, revolutionized modern science and, on the other, had deleterious impacts on biological systems. Cu^{1+} and Cu^{2+} are the two oxidation states in which copper is present. This feature enables copper to operate as an oxidizing or reducing agent in biological processes. However, it also renders copper potentially hazardous since Cu ions have the ability to promote the generation of reactive oxygen species and persuade oxidative stress [24]. Cu obtains remarkable qualities as a nanomaterial, such as its tiny size and large surface area, confer visual effects, magnetic, chemical reactivity, and physical resistance [25–27]. Due to their distinct characteristics, CuNP is used in many different sectors, such as electronics, paints, plastics, wastewater treatment, fungicides, chemicals, petrochemicals, textiles, electronics, cosmetics, and environmental remediation [28]. They have interacted with agriculture and the environment due to their considerable synthesis and application. However, the CuNPs become hazardous at a certain threshold, depending on the kind of crops and plants [2]. Since so few studies demonstrate the impact of CuNPs on agricultural plants, the effect of CuNPs on plants must be investigated.

2.1 Synthesis of CuNPs

CuNPs can be produced using specialized methods such as physical, chemical, and biological approaches. The prerequisites for synthesizing CuNPs and other metal nanoparticles include precursors for the generation of copper ions, reducing agents that may provide electrons for the reduction of copper ions, and nanoparticle aggregation.

2.1.1 Physical Method

To produce CuNPs, the necessary electrons for the reduction of copper to its elemental form in the range of 1–100 nm, this approach uses electric current as a source of electrons [29]. The physical strategy of synthesizing CuNPs often employs techniques such as laser pyrolysis, electro-spraying, high-energy ball milling, laser ablation, and evaporation–condensation [30]. Physical approaches are more advantageous than chemical methods due to the uniform dispersion of nanoparticles and the lack of reagent contamination [31].

2.1.2 Chemical Method

This approach reduces metal ions to their elemental form in sizes between 1 and 100 nm by using chemical compounds as the source of electrons [29]. Whereas certain harmful synthetic compounds like sodium borohydride and hypophosphite are utilized as reducing agents in traditional chemical procedures, naturally occurring substances like ascorbic acid and citric acid are employed in the green chemical approach [32]. The main drawbacks of this approach are its toxicity, environmental contamination, and operational complexity.

2.1.3 Biological Method

This process uses living things and their extracts (plants, algae, fungi, and bacteria) as a source of electrons to reduce copper salt and stabilize CuNPs [32–34]. The green-mediated technique for producing CuNPs has greatly benefited the fields of nanotechnology and nanoscience since organisms that are utilized as stabilizing and reducing agents are easily obtainable. The synthesis of Cu-nanoparticles from different biogenic sources is presented in Table 1. Studies have been conducted and verified about the efficiency of microorganisms and plants in lowering and maintaining copper ion levels. The primary issues with the physical and chemical techniques are their high cost and the use of hazardous and exclusive substances [35, 36]. The chemical approaches have a limited yield, are expensive, and are not environmentally friendly. Green synthesis, or biological approaches, are more favorable than physical and chemical ones since they are high-yielding, economical, and safe for the environment [36]. Sustainability issues serve as a major driving force behind green synthesis, which aims to create a contamination-free, green approach to the production of nanoparticles. The processes used to produce CuNPs via green synthesis are depicted in Fig. 1.

3 Impact of CuNPs on Plants

3.1 Effects of CuNPs on Seed Germination

There are several factors that influence how nanoparticles affect plants, including the type of plant being studied, size, concentration, composition, and physio-chemical characteristics of the NPs. The germination of seeds marks the start of a physiological process. The seed coat protects the embryo. The effect of CuNPs on plant growth has been depicted in Table 2 and Fig. 2. The radicle is the first tissue to come into direct contact with metals when the seed coat is corroded. Kausar et al. (2022) synthesized CuNP using mint extract and found that a 25 mg/L CuNP dose improved the germination and growth of wheat. Gomes et al. [54] noted that by promoting shoot growth and photosynthetic electron transfer, bulk Cu and chitosan-encapsulated Cu (NPCu) positively affected the early development of maize seedlings under favorable conditions. Gomes et al. [54] noted that by promoting shoot growth and photosynthetic electron transfer, bulk Cu and chitosan-encapsulated Cu (NPCu) positively affected the early development of maize seedlings under favorable conditions. However, under stress conditions caused by the seed deterioration process, NPCu was better than bulk Cu in activating the enzymatic defense system. Cu enhances plant disease resistance by activating the plant defense enzymes: polyphenol oxidases and phenylalanine ammonia-lyase [55]. A study with Cu-chitosan NPs on maize observed that maize seedlings' growth was improved by enhancing α-amylase activity and the starch content in germinating seedlings [56].

Table1 Different biological organisms used for the synthesis of copper nanoparticles

Types of biological source	Name of the species	Parts used	Salt and concentration	Size	References
Plants	*Solanum tuberosum*	Starch extract	$Cu(NO_3)_2$ (0.1 M)	54 nm	[37]
	Citrus medica	Fruit extract	$CuSO_4$ (1 mM)	54 nm	[38]
	Camellia sinensis/or Coffea arabica	Tea leaf extract/or coffee powder extract	$Cu(NO_3)_2$	50–100 nm	[39]
	Citrus medica Linn	Fruit juice	$CuSO_4$ (100 mM)	33 nm	Shende et al. (2015)
	Vitis vinifera	Leaf extract	$CuSO_4$ (1 mM)	-	[40]
	Centella asiatica	Leaf extract	Cu $(CH_3COO)_2$ (1% v/v)	2–5 μm	[41]
Bacteria	*Morganella morganii*		$CuSO_4$ (5 mM)	19.2 nm	[42]
	Pseudomonas fluorescens		$CuSO_4$ (318, 750, 1000 mg/L)	49 nm	[43]
	Hypocrea lixii		$CuCl_2$ (100 mg/L)	24.5 nm	Salvadori et al. (2013)
	Salmonella typhimurium		Cu $(NO_3)_2$ (1 mM)	49 nm	[44]
Fungi	*Stereum hirsutum*		$CuCl_2$, $Cu(NO_3)_2$ and $CuSO_4$ (5 mM),	5–20 nm	[45]
	Aspergillus niger strain STA9	Fungal culture extract	$CuSO_4$ (1, 3, 5, 10, 15, and 20 mM)	500 nm	[46]
	Shizophyllum commune	Fungal mycelia	$CuCl_2{\cdot}2H_2O$ (1 mM)	92 nm	[47]
	Stereum hirsutum	Mycelium-free extract	$CuCl_2$, $Cu(NO_3)_2$, and $CuSO_4$ (5 mM)	250–300 nm	[45]
	Aspergillus flavus	Fungal mycelia	$CuSO_4.5H_2O$ (5 mM)	2–60 nm	[48]
	Trichoderma asperellum	Cell-free extract	Cu $(NO_3)_2$. $3H_2O$ (5 mM)	110 nm	[49]
Algae	*Botryococcus braunii*	Algal extract	Cu $(CH_3COO)_2$ (1 mM	10–70 nm	[50]
	Anabaena cylindrica	Algal extract	$CuSO_4.5H_2O$ (1 mM)	3.6–43 nm	[51]
	Pterocladia capillacea	Algal extract	$CuSO_4$ (3 mM)	62 nm	[52]
	Chlorella kessleri	Culture solution	$CuSO_4$ (0.5 mM)	*15–45 nm*	[53]

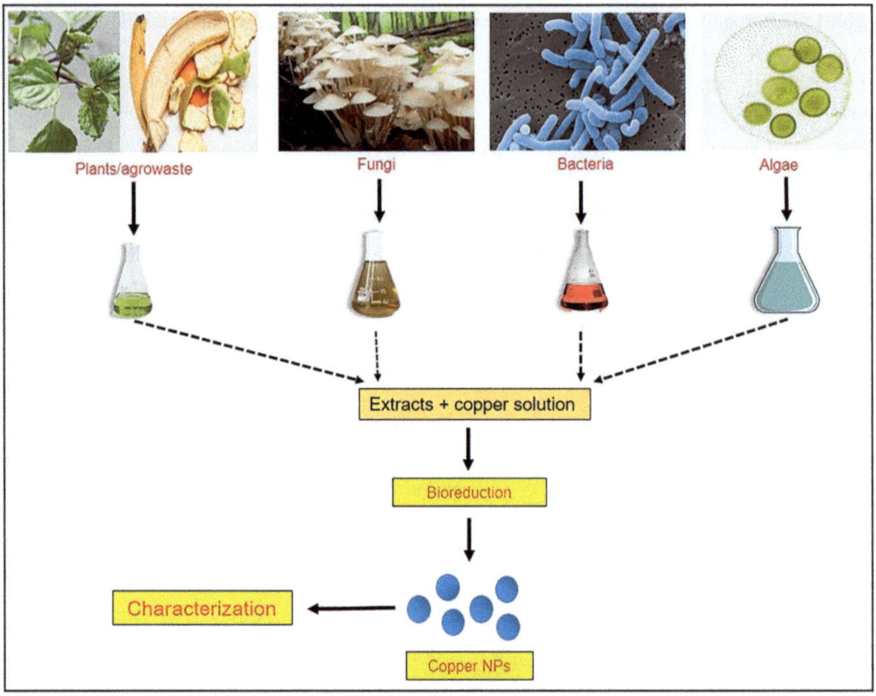

Fig. 1 Biosynthesis of CuNPs through different biological sources

Zafar et al. [76] studied the effect of CuO-NPs (1–20 mg/L) on mustard (*Brassica nigra*) seed germination and seedling growth. CuO is the most prevalent type of Cu at the plant root region. A hydroponic investigation involving rice (*Oryza sativa*) by [77] found that CuO-NPs could be taken by root and moved into the epidermis, exodermis, cortex, and endodermis but could not easily pass the Casparian strip, however, the growth of lateral roots provided an alternative route for CuO-NPs to get past the stele. Some studies demonstrate that NPs absorbed by roots are not migrated to shoots and may be localized in the epidermis and exodermis, or the process is not well-known [78]. An investigation done by [62] on copper nanoparticles/compounds found that CuO-NPs and Cu-compounds inhibited over 50% seed germination in Coriandrum sativum and impaired the nutritional value at 80 mg/kg in spiked soil. In the same experiment, however, the germination rate of *Oryza sativa* seeds was barely affected (7%) at 1000 mg/L of CuO-NPs compared to the control. In a study by Rajput et al. [79], CuO-NPs (10,000 mg/L) slowed the rate of germination and hampered the root and shoot growth of spring barley (*Hordeum sativum distichum*).

Table 2 Effect of copper nanoparticles on crop plants

Effects on plants	Crops	Concentration (mg/L)	Effects	References
Negative effect	Alfaalfa	0–20	Reduced the size of the plants, altered nutrient elements and enzymatic activity	[25–27]
	Wheat	200	Decreased more than 13% shoot and 59% roots length compared to untreated plants, and brown necrotic lesions appeared on the root	[57]
	Mungbean	0–500	Affected chlorophyll production, nutrient availability, decreased root and lateral root growth, and increased proline content	[58, 59]
	Maize	2–100	Chlorotic symptoms appeared on the root	[60]
	Cucumber	100–600	Inhibited seed germination (23.3%)	[61]
	Cilantro	0–80	Affected germination rate and shoot elongation	[62]
	Rice	0–1000	Decreased the number of thylakoids per granum	[63]
	Spinach	1000	Elevated concentration-affected plants, reduced the root and shoot length, total weight, and chlorophyll contents	[64])
	Onion	0–80	Reduced/stopped root growth at high concentration, deformed the surface of root cap and meristematic zone	[65]
	Mustard	0–1500	Observed thinner and brittle root with brown necrotic lesions. Affected ROS enzyme system, antioxidants, proline, lipid peroxidation	[66]
	Tomato	0–500	Reduced total chlorophyll content	[67]
	Soybean	50–500	Affected shoot growth, weight, and chlorophyll content as well as root length and its fresh weight	[67]

(continued)

Table 2 (continued)

Effects on plants	Crops	Concentration (mg/L)	Effects	References
	Carrot	1–1000	Reduced the shoots biomass and restricted the Cu accumulation in the taproot periderm	[58]
	Sweet potato	100–1000	Affected tuber biomass	[20]
	Barley		Induced the release of ROS, damaged cell membrane, decreased shoot and root lengths, and reduced the enzymatic activities	[68]
	Cotton	1000	Affected root length, aggregates observed on outer epidermis of root	[69]
	Zucchini	0–1000	Reduced 90% biomass	[70]
Positive effect	Wheat	50 and 100	Foliar spray increased growth parameters. Foliar spray is more effective than soaking method	[71]
	Barley	10, 25, 50, 100, 250	Enhancement seed germination and seedling growth	[72]
	Cabbage	70	Decreased plant weight, water content and photosynthesis. However increased root and shoot growth	[73]
	Chinese cabbage	10	Concentration-dependent response. In high amount CuO-NPs induced oxidative stress	[74]
	Soybean	5, 15, 30, 45, 60, 100	Concentration-dependent effect on root and shoot growth	[75]
	Chickpea	5, 15, 30, 45, 60	Low concentrations of CuO-NPs increased Growth parameters, while higher amounts Inhibited plant growth	[75]

3.2 *Effects of CuNPs on Plant Growth*

Copper, an essential element found in many proteins and enzymes, plays a vital role in plant health and nutrition, including chlorophyll biosynthesis and seed development

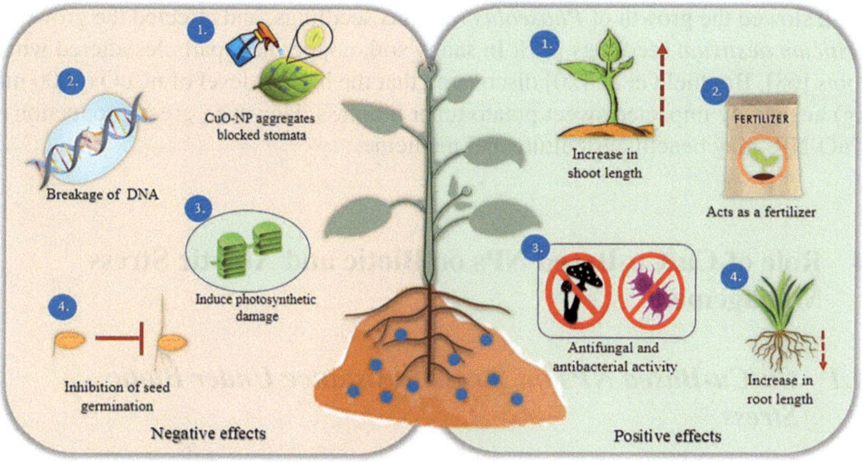

Fig. 2 Effect of CuNPs on plant germination and growth

[80]. According to [81], several significant crops, such as maize, are susceptible to copper deficiency. Additionally, Cu is a component of multiple enzyme systems that control the pace of numerous plant biochemical processes. Cu is also necessary for photosynthesis, respiration, protein, and carbohydrate metabolism (Ambrosini et al. 2018). The use of CuO-NPs, has been found to impact soybean drought tolerance as well as seed quality and yield (Tran et al. 2020). It is also well-known that as a micronutrient in low concentrations, Cu can promote plant growth and development [82]. However, [58] showed that high doses of CuO-NP negatively affect the uptake of essential micronutrients (B, Mn, Zn, Fe, Mo, and Mg) in soybean plants. The effect of different CuO-NPs on plants is depicted in Table 1. Olchowik et al. [83] observed that foliar application of CuO-NPs accumulated in the leaf and translocated to different parts. Foliar application of CuO-NPs on leaves causes the deposition of CuO-NP aggregates on the leaf surface, and these aggregates block and deform the stomatal apertures and produce abiotic stress in plants [73]. Singh and Kumar [64] applied CuNPs through irrigation water and observed that CuNPs reduced the root and shoot growth of spinach. According to [65], the CuO-NPs impeded the growth of the tip of the onion root and, after 72 h of exposure, totally halted the root's growth when treated with 80 mg/L CuO-NPs. CuNPs/compounds decreased the shoot length in Coriandrum sativum cultivated in soil rich in organic matter content [62]. Similar results were obtained by hydroponically grown alfalfa (*Medicago sativa*), as shown by Hong et al. [25–27]. Nair and Chung [84] discovered that CuO-NPs inhibited the development of Indian mustard (*Brassica juncea* L.), decreased the amount of total chlorophyll and carotenoid content, and shortened the main and lateral roots. CuO-NP accumulation caused an alteration in the ultrastructure of leaves, particularly in the photosynthetic apparatus. This was manifested by a decrease in the number of thylakoids per granum, plastoglobules, starch content, and stomatal aperture [83]. CuNP exposure is said to have decreased 90% of the biomass in *Cucurbita pepo*

[70], slowed the growth of *Phaseolus radiatus* seedlings, and affected the growth of *Triticum aestivum* seedlings [85]. In sandy soil, copper nanoparticles altered wheat roots [68]. Bradfield et al. [20] discovered that the highest level of nCuO (1000 mg/kg) negatively impacted sweet potato tuber biomass. However, green production of CuO-NPs may benefit agriculture and medicine.

4 Role of Cu/Cu-Based NPs on Biotic and Abiotic Stress Management

4.1 Cu/Cu-Based NPs on Stress Resistance Under Biotic Stress

The commercially available pesticide Kocide-3000 contains $Cu(OH)_2$NPs with well-established antibacterial and antifungal properties. CuNPs have antibacterial activity against plant fungal infections (*Fusarium oxysporum, Cochliobolus lunatus*, and *Alternaria alternata*) [86], as well as bacterial pathogens (*Escherichia coli* and *Bacillus subtilis*) [87]. CuNPs outperformed the conventional fungicide bavistin (Devistin, carbendazim 50% WP) in a research that compared their antibacterial efficacy. Cu-based nanomaterials (CuO-NPs and $Cu_3(PO_4)_2{\cdot}3H_2O$ nanosheets) have been compared recently by [88] for their capacity to control root fungal disease in watermelon (*Citrullus lanatus*) caused by *Fusarium oxysporum f. sp. niveum*. In a greenhouse study, Cu_3 $(PO_4)_2{\cdot}3H_2O$ nanosheets at a dose of 10 mg/L effectively inhibited fungal disease (measured by yield and a 58% reduction in disease development). CuO-NPs only significantly impacted disease at a considerably higher dose (1000 mg/L), whereas $Cu_3(PO_4)_2{\cdot}3H_2O$ nanosheets were more effective at lower dose. It is necessary to look at the possible toxicity from the high concentration (1000 mg/L). Lately, [89] examined the impact of CuNPs and potassium silicate NPs on tomato plants grown hydroponically and subjected to the pathogen *Clavibacter michiganensis*. According to the obtained data, both NPs effectively reduced the severity of the *C. michiganensis* infection. Additionally, researchers observed that the two combined NPs improved *C. michiganensis* tolerance by increasing the levels of enzymatic and non-enzymatic metabolic products crucial for tomato plants' defense. It has also been shown that CuNPs have insecticidal activity. The insecticidal activity of CuO-NPs on different types of cotton (conventional and Bt-transgenic) has been investigated by [90]. The NPs improved resistance by upregulating the expression of genes producing Bt toxin in cotton plant tissues at a concentration as low as 10 mg/L. CuO-NPs showed their entomotoxic properties against *Spodoptera littoralis*, indicating that these NPs might be used in economically viable pesticide formulations [91]. A pot and field-based study on *Curvularia* leaf spot (CLS) disease of maize observed that Cu-chitosan NPs at a dosage of 0.04–0.16% and 0.12–0.16% in pot and field study, respectively, have managed to control CLS disease [92].

4.2 Role of Cu/Cu-Based NPs on Stress Resistance Under Abiotic Stress

Biogenic CuNPs (25 and 50 mg/kg), which are produced by the bacterial strain *Klebsiella pneumonia*, as a soil amendment immobilizes chromium (Cr) in the soil and prevents its translocation in wheat [93]. The reduced translocation of Cr in above-ground biomass improves wheat growth by reducing cell oxidative stress. CuNP priming enhances the biomass and grain yield of maize plants under drought stress by increasing the amount of carotenoid content, chlorophyll, and leaf water content [80]. CuNPs priming also significantly regulate the amount of ROS in maize in response to drought by boosting antioxidant and ROS-scavenging enzyme activity. Farooq et al. [94] observed that seed-priming wheat with copper-chitosan nanoparticles (Cu-CNP) induces resistance to hyperosmotic stress and salinity induced by PEG.

5 Conclusion

The realm and utilization of nanomaterials are continuously expanding, with ongoing research efforts. Green synthesis methods, which inflict minimal to no harm on the environment and those involved in their creation, prove to be as effective as traditional synthesis methods. Considering these factors, it is imperative to approach research with environmental consciousness. Exploring the green synthesis of nanotechnology holds the potential to advance medical equipment, create innovative supercomputer conductors, and revolutionize sensor technology for space exploration. Our review delved into the effects and applications of copper nanoparticles, focusing primarily on the green synthesis of these nanoparticles, particularly concerning seed germination and plant growth. We have enlisted the methods of producing green nanomaterials, emphasizing the use of active molecules derived from various microorganism substrates to facilitate synthesis. Incorporating active molecules from natural biological systems continues to expand as more organisms become viable for utilization. As the world adapts to climate change, ongoing advancements in the green synthesis of nanomaterials become crucial for preserving environmental health, energy sustainability, and maintaining ethical standards in scientific research. While the toxicity of copper nanoparticles (CuNPs) has been widely reported in various plant species, positive findings are also documented.

Consequently, a meticulous determination of the environmental fate of CuO-NPs is essential, with defined criteria for sustainable applications. The increasing number of studies addressing the toxicity of CuO-NPs necessitates consideration alongside the exploration of new applications. Research has revealed the hazardous effects of NPs on germination, accumulation, and growth, underscoring the need for careful evaluation. Current and future research must be contextualized within the framework of existing risk assessments associated with CuO-NPs, considering their usage, distribution, and release into the environment.

6 Competing Interests

The authors declare that they have no competing interests.

7 Declaration of Generative AI and AI-Assisted Technologies in the Writing Process

During the preparation of this work the book chapter author(s) have not taken any assistance from AI, and the book chapter author(s) reviewed and edited the content as needed and take(s) full responsibility for the content of the publication. All authors read and approved the final version of the manuscript.

References

1. Bachheti RK, Bachheti A, Husen A (eds) (2024a) Metal and metal-oxide based nanomaterials: singapore: synthesis, agricultural, biomedical and environmental interventions. Springer Nature Singapore. https://doi.org/10.1007/978-981-99-7673-7
2. An YJ (2006) Assessment of comparative toxicities of lead and copper using plant assay. Chemosphere 62:1359–1365
3. Joshi A, Sharma A, Bachheti RK, Husen A, Mishra VK (2019) Plant-mediated synthesis of copper oxide nanoparticles and their biological applications. In: Nanomaterials and plant potential. Springer, Cham, pp 221–237. https://doi.org/10.1007/978-3-030-05569-1_8
4. Wassie AT, Ahmed IN, Bachheti A, Husen A, Bachheti RK (2024) Plant extract-assisted green synthesis and characterization of copper nanocomposite. In: Metal and metal-oxide based nanomaterials: synthesis, agricultural, biomedical and environmental interventions. Springer Nature Singapore, Singapore, pp 43–66. https://doi.org/10.1007/978-981-99-7673-7_3
5. Wegner K, Schimmoeller B, Thiebaut B, Fernandez C, Rao TN (2011) Pilot plants for industrial nanoparticle production by flame spray pyrolysis. Kona Powder Part J 29:251–265
6. Amendola V, Meneghetti M (2013) What controls the composition and the structure of nanomaterials generated by laser ablation in liquid solution? Phys Chem Chem Phys 15(9):3027–3046
7. Song H, Tang A, Xu G, Liu L, Yin M, Pan Y (2018) One-step convenient hydrothermal synthesis of MoS2/RGO as a high-performance anode for sodium-ion batteries. Int J Electrochem Sci 13(5):4720–4730
8. Walter JG, Petersen S, Stahl F, Scheper T, Barcikowski S (2010) Laser ablation-based one-step generation and bio-functionalization of gold nanoparticles conjugated with aptamers. J Nanobiotechnol 8(1):1–11
9. Aboulouard A, Gultekin B, Can M, Erol M, Jouaiti A, Elhadadi B, Demic S et al (2020) Dye sensitized solar cells based on titanium dioxide nanoparticles synthesized by flame spray pyrolysis and hydrothermal sol-gel methods: a comparative study on photovoltaic performances. J Market Res 9(2):1569–1577
10. Joshi DR, Adhikari N (2019) An overview on common organic solvents and their toxicity. J Pharmaceutical Res Int 28(3):1–18
11. Lin B, Zhu X, Fang L, Liu X, Li S, Zhai T, Xia H et al (2019) Birnessite nanosheet arrays with high K content as a high-capacity and ultrastable cathode for K-ion batteries. Adv Mater 31(24):1900060

12. Mueller R, Jossen R, Pratsinis SE, Watson M, Akhtar MK (2004) Zirconia nanoparticles made in spray flames at high production rates. J Am Ceram Soc 87(2):197–202
13. Huston M, DeBella M, DiBella M, Gupta A (2021) Green synthesis of nanomaterials. Nanomaterials 11(8):2130
14. Sivaraj A, Kumar V, Sunder R, Parthasarathy K, Kasivelu G (2020) Commercial yeast extracts mediated green synthesis of silver chloride nanoparticles and their anti-mycobacterial activity. J Cluster Sci 31:287–291
15. Saratale RG, Karuppusamy I, Saratale GD, Pugazhendhi A, Kumar G, Park Y, Shin HS et al (2018) A comprehensive review on green nanomaterials using biological systems: recent perception and their future applications. Colloids Surf B 170:20–35
16. Birla SS, Tiwari VV, Gade AK, Ingle AP, Yadav AP, Rai MK (2009) Fabrication of silver nanoparticles by Phoma glomerata and its combined effect against *Escherichia coli*, Pseudomonas aeruginosa and Staphylococcus aureus. Lett Appl Microbiol 48(2):173–179
17. Babu MG, Gunasekaran P (2009) Production and structural characterization of crystalline silver nanoparticles from Bacillus cereus isolate. Colloids Surf B 74(1):191–195
18. Bansal V, Rautaray D, Ahmad A, Sastry M (2004) Biosynthesis of zirconia nanoparticles using the fungus Fusarium oxysporum. J Mater Chem 14(22):3303–3305
19. Philip D (2009) Biosynthesis of Au, Ag and Au–Ag nanoparticles using edible mushroom extract. Spectrochim Acta Part A Mol Biomol Spectrosc 73(2):374–381
20. Bradfield SJ, Kumar P, White JC, Ebbs SD (2017) Zinc, copper, or cerium accumulation from metal oxide nanoparticles or ions in sweet potato: yield effects and projected dietary intake from consumption. Plant Physiol Biochem 110:128–137
21. Chibber S, Ansari SA, Satar R (2013) New vision to CuO, ZnO, and TiO_2 nanoparticles: their outcome and effects. J Nanopart Res 15:1–13
22. Nekrasova GF, Ushakova OS, Ermakov AE, Uimin MA, Byzov IV (2011) Effects of copper (II) ions and copper oxide nanoparticles on Elodea densa Planch. Russ J Ecol 42(6):458–463. https://doi.org/10.1134/S1067413611060117
23. Baker DE, Senef JP (1995) Copper. In: Alloy BJ (ed) Heavy metals in soils. Blackie Academic and Professional, London, pp 179–205
24. Hänsch R, Mendel RR (2009) Physiological functions of mineral micronutrients (Cu, Zn, Mn, Fe, Ni, Mo, B, Cl). Curr Opin Plant Biol 12:259–266
25. Hong J, Rico C, Zhao L, Adeleye AS, Keller AA, Peralta-Videa JR et al (2015) Toxic effects of copper-based nanoparticles or compounds to lettuce (*Lactuca sativa*) and alfalfa (*Medicago sativa*). Environ Sci Proc Imp 17:177–185. https://doi.org/10.1039/C4EM00551A. PMID: 25474419
26. Hong J, Rico CM, Zhao L, Adeleye AS, Keller AA, Peralta-Videa JR, Gardea-Torresdey JL (2015) Toxic effects of copper-based nanoparticles or compounds to lettuce (*Lactuca sativa*) and alfalfa (*Medicago sativa*). Environ Sci Processes Impacts 17:177–185
27. Hong J, Rico CM, Zhao L, Adeleye AS, Keller AA, Peralta-Videa JR, Gardea-Torresdey JL (2015) Toxic effects of copper-based nanoparticles or compounds to lettuce (*Lactuca sativa*) and alfalfa (*Medicago sativa*). Environ Sci Processes Impacts 17:177–185
28. Rafique M, Shaikh AJ, Rasheed R, Tahir MB, Bakhat HF, Rafique MS, Rabbani F (2017) A review on synthesis, characterization and applications of copper nanoparticles using green method. NANO 12:04
29. Hussain I, Singh NB, Singh A, Singh H, Singh SC (2016) Green synthesis of nanoparticles and its potential application. Biotechnol Lett 38(4):545–560
30. Kuang L, Mitchell BS, Fink MJ (2015) Silicon nanoparticles synthesised through reactive high-energy ball milling: enhancement of optical properties from the removal of iron impurities. J Exp Nanosci 10(16):1214–1222
31. Hornyak GL, Dutta J, Tibbals HF, Rao A (2008) Introduction to nanoscience. CRC Press
32. Akintelu SA, Oyebamiji AK, Olugbeko SC, Latona DF (2021) Green chemistry approach towards the synthesis of copper nanoparticles and its potential applications as therapeutic agents and environmental control. Current Res Green Sustain Chem 4:100176

33. Bachheti RK, Bachheti A (eds) (2023) Secondary metabolites from medicinal plants: nanoparticles synthesis and their applications. CRC Press, Boca Raton, USA. https://doi.org/10.1201/9781003213727
34. Siddiqi KS, Husen A (2020) Current status of plant metabolite-based fabrication of copper/copper oxide nanoparticles and their applications: a review. Biomaterials Research 24(11):1–15. https://doi.org/10.1186/s40824-020-00188-1
35. Sarkar J, Mridha D, Davoodbasha MA, Banerjee J, Chanda S, Ray K, Roychowdhury T, Acharya K, Sarkar J (2023) A state-of-the-art systemic review on selenium nanoparticles: mechanisms and factors influencing biogenesis and its potential applications. In: Biological trace element research, pp 1–37
36. Sarkar J, Mridha D, Sarkar J, Orasugh JT, Gangopadhyay B, Chattopadhyay D, Roychowdhury T, Acharya K (2021) Synthesis of nanosilica from agricultural wastes and its multifaceted applications: a review. Biocatal Agric Biotechnol 37:102175
37. Alishah H, Pourseyedi S, Ebrahimipour SY, Mahani SE, Rafiei N (2016) Green synthesis of starch-ediated CuO nanoparticles: preparation, characterization, antimicrobial activities and in vitro MTT assay against MCF-7 cell line. Rendiconti Lincei. https://doi.org/10.1007/s12210-016-0574-y
38. Nagaonkar D, Shende S, Rai M (2015) Biosynthesis of copper nanoparticles and its effect on actively dividing cells of mitosis in *Allium cepa*. Biotechnol Prog 31(2):557–565. https://doi.org/10.1002/btpr.2040
39. Sutradhar P, Saha M, Maiti D (2014) Microwave synthesis of copper oxide nanoparticles using tea leaf and coffee powder extracts and its antibacterial activity. J Nanostruct Chem 4:86. https://doi.org/10.1007/s40097-014-0086-1
40. Angrasan JKVM, Subbaiya R (2014) Biosynthesis of copper nanoparticles by Vitis vinifera leaf aqueous extract and its antibacterial activity. Int J Curr Microbiol Appl Sci 3(9):768–774
41. Devi HS, Singh TD (2014) Synthesis of copper oxide nanoparticles by a novel method and its application in the degradation of methyl orange. Adv Electron Electr Eng 4(1):83–88. https://doi.org/10.2174/0929867054637617
42. Ramanathan R, Field MR, O'Mullane AP, Smooker PM, Bhargava SK, Bansal V (2013) Aqueous phase synthesis of copper nanoparticles: a link between heavy metal resistance and nanoparticle synthesis ability in bacterial systems. Nanoscale 5:2300–2306. https://doi.org/10.1039/c2nr32887a
43. Shantkriti S, Rani P (2014) Biological synthesis of copper nanoparticlesusing *Pseudomonas fluorescens*. Int J Curr Microbiol App Sci 3(9):374–383
44. Ghorbani HR, Mehr FP, Poor AK (2015) Extracellular synthesis of copper nanoparticles using culture supernatants of *Salmonella typhimurium*. Orient J Chem 31(1):527–529. https://doi.org/10.13005/ojc/310165
45. Cuevas R, Dura´n N, Diez MC, Tortella GR, Rubilar O (2015) Extracellular biosynthesis of copper and copper oxide nanoparticles by *Stereum hirsutum*, a native white-rot fungus from Chilean forests. J Nanomater. https://doi.org/10.1155/2015/789089
46. Noor S, Shah Z, Javed A, Ali A, Hussain SB, Zafar S, Ali H, Muhammad SA (2020) A fungal based synthesis method for copper nanoparticles with the determination of anticancer, antidiabetic and antibacterial activities. J Microbiol Methods 174:105966
47. Fatima F, Wahid I (2022) Eco-friendly synthesis of silver and copper nanoparticles by *Shizophyllum commune* fungus and its biomedical applications. Int J Environ Sci Technol 19:7915–7926. https://doi.org/10.1007/s13762-021-03517-6
48. Saitawadekar A, Kakde UB (2020) Green synthesis of copper nanoparticles using *Aspergillus flavus*. J Crit Rev 7(9)
49. Saravanakumar K, Shanmugam S, Varukattu NB, MubarakAli D, Kathiresan K, Wang MH (2019) Biosynthesis and characterization of copper oxide nanoparticles from indigenous fungi and its effect of photothermolysis on human lung carcinoma. J Photochem Photobiol B 190:103–109
50. Arya A, Gupta K, Chundawat TS, Vaya D (2018) Biogenic synthesis of copper and silver nanoparticles using green alga Botryococcus braunii and its antimicrobial activity. In: Bioinorganic chemistry and applications

51. Bhattacharya P, Swarnakar S, Ghosh S, Majumdar S, Banerjee S (2019) Disinfection of drinking water via algae mediated green synthesized copper oxide nanoparticles and its toxicity evaluation. J Environ Chem Eng 7(1):102867
52. Aboeita NM, Fahmy SA, El-Sayed MM, Azzazy HMES, Shoeib T (2022) Enhanced anti-cancer activity of nedaplatin loaded onto copper nanoparticles synthesized using red algae. Pharmaceutics 14(2):418
53. Salas-Herrera G, González-Morales S, Benavides-Mendoza A et al (2019) Impact of microalgae culture conditions over the capacity of copper nanoparticle biosynthesis. J Appl Phycol 31:2437–2447. https://doi.org/10.1007/s10811-019-1747-8
54. Gomes DG, Pelegrino MT, Ferreira AS, Bazzo JH, Zucareli C, Seabra AB, Oliveira HC (2021) Seed priming with copper-loaded chitosan nanoparticles promotes early growth and enzymatic antioxidant defense of maize (*Zea mays* L.) seedlings. J Chem Technol Biotechnol 96(8):2176–2184
55. Servin A, Elmer W, Mukherjee A, De la Torre-Roche R, Hamdi H, White JC, Bindraban P, Dimkpa C (2015) A review of the use of engineered nanomaterials to suppress plant disease and enhance crop yield. J Nanopart Res 17(2):92
56. Saharan V, Kumaraswamy RV, Choudhary RC, Kumari S, Pal A, Raliya R, Biswas P (2016) Cu-chitosan nanoparticle mediated sustainable approach to enhance seedling growth in maize by mobilizing reserved food. J Agric Food Chem 64(31):6148–6155
57. Dimkpa CO, McLean JE, Latta DE, Manangón E, Britt DW, Johnson WP (2012) CuO and ZnO nanoparticles: phytotoxicity, metal speciation, and induction of oxidative stress in sandgrown wheat. J Nanopart Res 14:1125
58. Nair PG, Chung I (2014) A mechanistic study on the toxic effect of copper oxide nanoparticles in soybean (*Glycine max* L.) root development and lignification of root cells. Biol Trace Elem Res 162:342–352
59. Nair PMG, Chung M (2014) Copper oxide nanoparticle toxicity in mungbean (*Vigna radiata* L.) seedlings: physiological and molecular level responses of in vitro grown plants. Acta Physiol Plant 36:2947–2958
60. Wang Z, Xie X, Zhao J, Liu X, Feng W, White JC, Xing B (2012) Xylem- and phloem-based transport of CuO nanoparticles in maize (*Zea mays* L.) Environ Sci Technol 46:4434–4441
61. Moon YS, Park ES, Kim TO, Lee HS, Lee SE (2014) SELDI-TOF MS-based discovery of a biomarker in *Cucumis sativus* seeds exposed to CuO nanoparticles. Environ Toxicol Phar 38:922–931
62. Zuverza-Mena N, Medina-Velo IA, Barrios AC, Tan W, Peralta-Videa JR, Gardea-Torresdey JL (2015) Copper nanoparticles/compounds impact agronomic and physiological parameters in cilantro (*Coriandrum sativum*). Environ Sci Process Impacts 17:1783–1793
63. Costa DMVJ, Sharma PK (2016) Effect of copper oxide nanoparticles on growth, morphology, photosynthesis, and antioxidant response in Oryza sativa. Photosynthetica 54:110
64. Singh D, Kumar A (2016) Impact of irrigation using water containing CuO and ZnO nanoparticles on *Spinach oleracea* grown in soil media. Bull Environ Contam Toxicol 97:548–553
65. Deng F, Wang S, Xin H (2016) Toxicity of CuO nanoparticles to structure and metabolic activity of Allium cepa root tips. Bull Environ Contam Toxicol 97:702–708
66. Rao S, Shekhawat GS (2016) Phytotoxicity and oxidative stress perspective of two selected nanoparticles in Brassica juncea. 3. Biotech 6:244
67. Singh A, Singh NB, Hussain I, Singh H (2017) Effect of biologically synthesized copper oxide nanoparticles on metabolism and antioxidant activity to the crop plants *Solanum lycopersicum* and *Brassica oleracea* var. botrytis. J Biotechnol 262:11–27
68. Shaw AK, Ghosh S, Kalaji HM, Bosa K, Brestic M, Zivcak M, Hossain Z (2014) Nano-CuO stress induced modulation of antioxidative defense and photosynthetic performance of syrian barley (*Hordeum vulgare* L.) Environ Experimental Botany 102:37–47
69. Nhan LV, Yukui R, Weidong C, Jianying S, Shutong L, Trung NQ, Liming L (2016) Toxicity and bio-effects of CuO nanoparticles on transgenic Ipt-cotton. J Plant Interactions 11:108–116

70. Stampoulis D, Sinha SK, White JC (2009) Assaydependent phytotoxicity of nanoparticles to plants. Environ Sci Technol 43:9473–9479
71. Badawy AA, Abdelfattah NA, Salem SS, Awad MF, Fouda A (2021) Efficacy assessment of biosynthesized copper oxide nanoparticles (CuO-NPs) on stored grain insects and their impacts on morphological and physiological traits of wheat (*Triticum aestivum* L.) plant. Biology 10(3):233. https://doi.org/10.3390/biology10030233
72. Kadri O, Karmous I, Kharbech O, Arfaoui H, Chaoui A (2022) Cu and CuO nanoparticles affected the germination and the growth of barley (*Hordeum vulgare* L.) seedling. Bull Environ Contam Toxicol 108(3):585–593. https://doi.org/10.1007/s00128-021-03425-y
73. Xiong T, Dumat C, Dappe V, Vezin H, Schreck E, Shahid M, Sobanska S (2017) Copper oxide nanoparticle foliar uptake, phytotoxicity, and consequences for sustainable urban agriculture. Environ Sci Technol 51(9):5242–5251. https://doi.org/10.1021/acs.est.6b05546
74. Wang W, Ren Y, He J, Zhang L, Wang X, Cui Z (2020) Impact of copper oxide nanoparticles on the germination, seedling growth, and physiological responses in *Brassica pekinensis* L. Env Sci Pollut Res 27:31505–31515. https://doi.org/10.1007/s11356-020-09338-3
75. Adhikari T, Kundu S, Biswas AK, Tarafdar JC, Rao AS (2012) Effect of copper oxide nano particle on seed germination of selected crops. J Agr Sci Technol A 2(6A):815
76. Zafar H, Ali A, Zia M (2017) CuO nanoparticles inhibited root growth from *Brassica nigra* seedlings but induced root from stem and leaf explants. Appl Biochem Biotechnol 181:365–378
77. Peng C, Duan D, Xu C, Chen Y, Sun L, Zhang H, Yuan X, Zheng L, Yang Y, Yang J, Zhen X, Chen Y, Shi J (2015) Translocation and biotransformation of CuO nanoparticles in rice (*Oryza sativa* L.) plants. Environ Pollut 197:99–107
78. Miralles P, Church TL, Harris AT (2012) Toxicity, uptake, and translocation of engineered nanomaterials in vascular plants. Environ Sci Technol 46:9224–9239
79. Rajput VD, Minkina T, Mandzhieva S, Duply N, Fedorenko A, Sushkova S, Tsitsuashvili V (2017) Influence of copper oxide nanoparticle on seed germination and seedling growth. In: International scientific conference on Bmodern technologies in the study of biodiversity and planting of plants, Rostov, Russia, pp 115–116. http://conference2017.bg.sfedu.ru/index.php?option=com_content&view=article&id=12&Itemid=129. Accessed 26 Oct 2017
80. Van Nguyen D, Nguyen HM, Le NT, Nguyen KH, Nguyen HT, Le HM, Nguyen AT, Dinh NTT, Hoang SA, Van Ha C (2021) Copper nanoparticle application enhances plant growth and grain yield in maize under drought stress conditions. J Plant Growth Regulation 1–12
81. Karamanos RE, Pomarenski Q, Goh TB, Flore NA (2004) The effect of foliar copper application on grain yield and quality of wheat. Can J Plant Sci 84:47–56. https://doi.org/10.4141/P03-090
82. Karlsson HL, Gustafsson J, Cronholm P, Möller L (2009) Size-dependent toxicity of metal oxide particles—a comparison between nano- and micrometer size. Toxicol Lett 188:112–118
83. Olchowik J, Bzdyk RM, Studnicki M, Bederska-Błaszczyk M, Urban A, Aleksandrowicz-Trzcińska M (2017) The effect of silver and copper nanoparticles on the condition of English oak (*Quercus robur* L.) seedlings in a container nursery experiment. Forests 8:310
84. Nair PMG, Chung IM (2015) Study on the correlation between copper oxide nanoparticles induced growth suppression and enhanced lignification in indian mustard (*Brassica juncea* L.) Ecotoxicol Environ Safety 113:302–313
85. Lee WM, An YJ, Yoon H, Kweon HS (2008) Toxicity and bioavailability of copper nanoparticles to the terrestrial plants mungbean (*Phaseolus radiatus*) and wheat (*Triticum aestivum*): plant agar test for water-insoluble nanoparticles. Environ Toxicol Chem 27:1915–1921
86. Kanhed P, Birla S, Gaikwad S, Gade A, Seabra AB, Rubilar O, Duran N, Rai M (2014) In vitro antifungal efficacy of copper nanoparticles against selected crop pathogenic fungi. Mater Lett 115:13–17
87. Yoon K-Y, Hoon Byeon J, Park J-H, Hwang J (2007) Susceptibility constants of *Escherichia coli* and *Bacillus subtilis* to silver and copper nanoparticles. Sci Total Environ 373(2):572–575
88. Borgatta J, Ma C, Hudson-Smith N, Elmer W, Plaza Pérez CD, De La Torre-Roche R, Zuverza-Mena N, Haynes CL, White JC, Hamers RJ (2018) Copper based nanomaterials suppress root fungal disease in watermelon (*Citrullus lanatus*): role of particle morphology, composition and dissolution behavior. ACS Sustain Chem Eng 6(11):14847–14856

89. Cumplido-Nájera CF, González-Morales S, Ortega-Ortíz H, Cadenas-Pliego G, Benavides-Mendoza A, Juárez-Maldonado A (2019) The application of copper nanoparticles and potassium silicate stimulate the tolerance to Clavibacter michiganensis in tomato plants. Sci Hortic 245:82–89
90. Le Van N, Ma C, Shang J, Rui Y, Liu S, Xing B (2016) Effects of CuO nanoparticles on insecticidal activity and phytotoxicity in conventional and transgenic cotton. Chemosphere 144:661–670
91. Ayoub HA, Khairy M, Elsaid S, Rashwan FA, Abdel-Hafez HF (2018) Pesticidal activity of nanostructured metal oxides for generation of alternative pesticide formulations. J Agric Food Chem 66(22):5491–5498
92. Choudhary RC, Kumaraswamy RV, Kumari S, Sharma SS, Pal A, Raliya R, Biswas P, Saharan V (2017) Cu-chitosan nanoparticle boost defense responses and plant growth in maize (*Zea mays* L.). Sci Rep 7(1):9754
93. Noman M, Shahid M, Ahmed T, Tahir M, Naqqash T, Muhammad S, Song F, Abid HMA, Aslam Z (2020) Green copper nanoparticles from a native *Klebsiella pneumoniae* strain alleviated oxidative stress impairment of wheat plants by reducing the chromium bioavailability and increasing the growth. Ecotoxicol Environ Saf 192:110303
94. Farooq T, Nisa ZU, Hameed A, Ahmed T, Hameed A (2022) Priming with copper-chitosan nanoparticles elicit tolerance against PEG-induced hyperosmotic stress and salinity in wheat. BMC Chem 16(1):23. https://doi.org/10.1186/s13065-022-00813-1

Copper-Based Nanomaterials for Control of Plant Pathogens and Diseases

Parul Singh, Shweta Sharma, Kundan Kumar Chaubey, Anish Kumar Pal, Minerva Sharma, Ashwani Kumar Sanghi, Faizanul Haque Nagrami, Uma Sharma, and Priyamvada

Abstract Insect pests, soil-borne phytopathogens, parasites, and predators are among the biotic elements that mainly cause crop losses. Food rots by different microorganisms account for most food losses in the food sector. The agricultural and food production of crop productivity increased with advanced nanotechnology-used nanomaterials. The antibacterial properties of copper/copper oxide nanowires against various pathogens are widely recognized. Among the essential micronutrients, copper is crucial for plant growth and development. Agriculture makes fertilizers and pesticides based on copper nanoparticles (CuNPs). Copper nanomaterials' (CuNMs) small size makes it easier for plants to absorb them. Growing interest has been shown in using nanotechnology in agriculture to increase agricultural yields. In particular, because of their antibacterial characteristics, copper-based nanostructures are great substitutes to reduce plant diseases. However, more research must be conducted on its advantages in managing plant diseases. The usage of nanostructures depends on plant pathogens controlled by copper, and diseases are covered in this chapter.

Keywords CuNP · Plant · Pathogen · Diseases · Cu nanomaterials · Insects · Control

P. Singh · S. Sharma · F. H. Nagrami · U. Sharma · Priyamvada
College of Biotechnology, DUVASU, Mathura, Uttar Pradesh, India

K. K. Chaubey (✉) · M. Sharma
Department of Biotechnology, School of Basic and Applied Sciences, Sanskriti University, Mathura, India
e-mail: kundan2006chaubey@gmail.com

A. K. Pal
Division of Research and Innovation, School of Applied and Life Sciences, Uttaranchal University, Dehradun, India

A. K. Sanghi
School of Allied and Health Sciences, MVN University, Palwal, Haryana, India

R. K. Bachheti et al. (eds.), *Copper-Based Nanomaterial for Agricultural Practices*, Smart Nanomaterials Technology, https://doi.org/10.1007/978-981-96-8610-0_9

1 Introduction

Nanomaterials (NMs) are materials characterized by at least one nanoscale dimension and a diameter not exceeding 100 nm. Their distinct attributes, encompassing chemical and physical traits, alongside a notable high surface-to-volume ratio, have driven a surge in their utilization across diverse industries [3–6, 26]. These encompass plastics, consumer goods, agriculture, electronics, coatings, cosmetics, chemicals, food and packaging, environmental remediation, fuel components, energy, textiles, paints, and pioneering medical applications ([3, 27–31, 33, 47, 48, 55, 56] Taghiyari, 2023; Siddiqi and Husen 2022; Choudhary et al. 2024). Many technological advancements in agriculture and chemical additives, referred to as amendments here, have been implemented to boost crop yields while requiring the least amount of extra resource input. These methods include genetic alteration, prescriptive planting, sensor-based crop quality monitoring, and the creation of novel agrichemical delivery systems [61]. Disease and pest-related crop losses continue to be a significant issue, resulting in 20–40% yield reductions and accompanying financial losses, endangering global food security.

Large surface area, copper-based NMs reactivity, and possible improved movement within plant tissues make them superior to conventional approaches [60, 61]. Moreover, material properties are being fine-tuned to tailor micronutrient release for maximum benefit, which is of great interest [49]. Studies on the application of copper nanomaterials (CuNMs) have been motivated by the metal ions and established antibacterial and antifungal characteristics [19]. In the past, bulk hydroxide or salt applications of copper have been used to suppress pathogens in crops (Weng et al. 1997). Despite being reasonably successful for the intended use, using solubilized salts of copper may cause copper to accumulate in the field and leach into freshwater sources. Their catalytic (mainly), optical and conductive capabilities led to their initial applications in energy conversion and storage, as well as chemical and material manufacture [22]. In the last few decades, five primary approaches have been developed to generate Cu-based NMs: electrochemical synthesis, photochemical reaction procedures, heat treatment, chemical treatment, and sonochemical treatment. Biological approaches have recently led to the development of a more inexpensive, ecological, waste-free, and effective technology [43]. The problem nowadays, excluding the technology, is to produce CuNMs in a tiny size range with high stability and antibacterial activity. A novel family of antimicrobials known as biogenic CuNMs has the ability to combat a variety of bacteria, especially those that have become resistant to antibiotics [63]. Rai et al. [58] remarked on this. However, a great deal remains to be discovered regarding the exact mechanism of action of CuNMs. As per [13], the prevailing theory suggests that NMs based on copper induce cellular death through modifications to the permeability of the cellular membrane. According to [72], NMs may also impede essential biological processes like DNA replication by absorbing their metallic ions.

A new study indicates that CuNMs may have a biocidal effect due to several mechanisms, including physical and chemical degradation and suppression of gene expression [14]. In the last twenty years, agricultural applications have witnessed a notable increase in the use of engineered nanomaterials. They are used for a broad range of tasks, including delivering nutrients, biological material, and pesticides; stabilizing biopesticides; and creating nano biosensors to identify diseases or pesticides that are chemical-based [1, 25]. Malandrakis et al. [44] have developed and evaluated several metal nanomaterials, such as copper, zinc, silver, magnesium, and titanium, for their antibacterial effectiveness in treating bacterial and fungal plant diseases. Most of the discovered nanoproducts have the potential to be an outstanding tool for reducing pesticide intake into the environment through low-dose application, as they assure efficacy against the studied targets that is equal to or greater than that of readily accessible conventional pesticides.

Plant diseases caused by bacteria and fungi have been controlled for almost a century with copper fungicides [68]. Both conventional and organic horticulture production currently uses a variety of copper compounds, including copper oxychloride, copper hydroxide, copper sulfate, copper carbonate, and copper oxide. Copper fungicides are widely used across the world, although they may be dangerous for the environment and non-target species. With the main focus being on lowering application doses, recent developments in nanotechnology may offer ways to get around restrictions on the overuse of copper fungicides. One of several metal nanomaterials that were developed and tested for their antibacterial efficiency, CuNMs became the primary weapon due to their broad availability and lower manufacturing costs [58]. A multitude of current studies have demonstrated the significant potential of CuNMs, which can be manufactured in a variety of ways, to combat oomycetes and other plant-pathogenic fungi that cause illnesses and infect several crops [34, 51].

2 Nanomaterials Based on Copper to Regulate Microbiological Activity

According to [39], nano-agriculture is a cutting-edge field in contemporary farming that has significant potential to improve food quality and nutrition studies. Because they use fewer chemical resources, nano-agrochemicals have the potential to be leaders in this industry. Traditional fertilizers and pesticides have a significant ecological footprint since they release fewer chemicals into the environment. Numerous inorganic NMs have demonstrated effectiveness in nano-agricultural applications, including copper-based goods [24], aluminum [65], and silver [38]. According to [51], CuNMs specifically have proven to help protect plants from a variety of threats, such as *Fusarium sp.*, *Phoma destructiva*, *Fusarium oxysporum*, *Alternaria alternata*, *Penicillium italicum*, *Penicillium digitatum*, and *Rhizoctonia solani*. Research has shown that several types of CuNMs having size in 11–14 nm in diameter can reduce

Phytophthora infestans in smaller amounts (15–35 g/hl) have approved the use of Cu-based products (35–224 g/hl). Furthermore, a small amount of research has demonstrated CuNMs biocidal properties against bacterial phytopathogens, including *Pseudomonas spp.*, *Xanthomonas spp.*, and *Ralstonia solanacearum* [7, 14]. However, most studies have focused on their effectiveness against human pathogens like *E. coli, P. aeruginosa,* and *Klebsiella pneumonia,* among others [64]. Metallic oxide NMs such as those derived from Ag, Mg, Si, and carbon have demonstrated the ability to reduce economically significant plant bacteria like *Xanthomonas citri pv. citri* and *Xylella fastidiosa* [20]. The cost-effectiveness, protective capabilities, and reduced risk of resistance development due to their multi-site mode of action against pathogens have positioned bulk copper compounds as effective tools. The antimicrobial activity of chemically synthesized CuNMs was documented by [59] against both fungi and bacteria, including *Aspergillus niger*, *Aspergillus flavus*, *Candida albicans*, and *Micrococcus luteus*, as well as *S. aureus*, *E. coli*, *Klebsiella pneumoniae*, and *Pseudomonas aeruginosa*. The study reported CuNMs to be particularly effective against *E. coli, C. albicans, S. aureus, M. luteus, A. niger, K. pneumoniae, A. flavus,* and *P. aeruginosa* in that order [59]. Safeguarding agricultural crops from various pests, including those causing broad-spectrum bacterial and fungal infections, has been shown by [35]. According to [35], incorporating copper fungicides into disease management tactics in organic agriculture has been made easier by the residues' easy removal from plant tissues and lack of systemic mobility. Copper-containing nanomaterials, or CuNMs, are being used more and more in agricultural and food preservation due to their unique nanoscale qualities that come from antifungal and antibacterial qualities [35, 50]. The small size of NMs enables more effective penetration into microbial cells within the gradient, allowing for enhanced control at lower application dosages, as emphasized by Nel et al. in 2006. Although the mechanism of CuNMs antimicrobial action remains partially understood, this aligns with previous studies demonstrating their antimicrobial activity against various phytopathogens.

Currently, nanotechnology has proposed a variety of tools to address agricultural concerns. According to reports, the growth of NMs and nano-treatments for diseases of plants can increase crop yields through nanotechnology. [17, 69]. Indeed, the exploration of numerous nanomaterials for treating plant diseases has been a subject of research. Baker et al. [8] highlighted that CuNMs possess significant potential as effective fungicides. Synthesizing CuNMs in an economically viable manner for agricultural applications is crucial. Several synthesis techniques are available, including hydrothermal, microwave, photochemical, electrochemical, microemulsion, and chemical reduction methods. Selecting the appropriate synthesis method is pivotal for their successful utilization in agriculture. Parveen et al. [52] have explained many existing synthesis methods that involve expensive equipment, hazardous chemicals, organic solvents, and critical reaction conditions, so there is a notable need for a green synthesis approach. This method could be recommended for producing stable CuNMs by utilizing inexpensive and ecologically friendly ingredients. Such an approach aligns with sustainability goals and minimizes the environmental impact associated with nanoparticle synthesis. Here, we suggest a synthetic technique that leverages ascorbic acid to reduce Cu^{2+} ions. Additionally, ascorbic acid widely

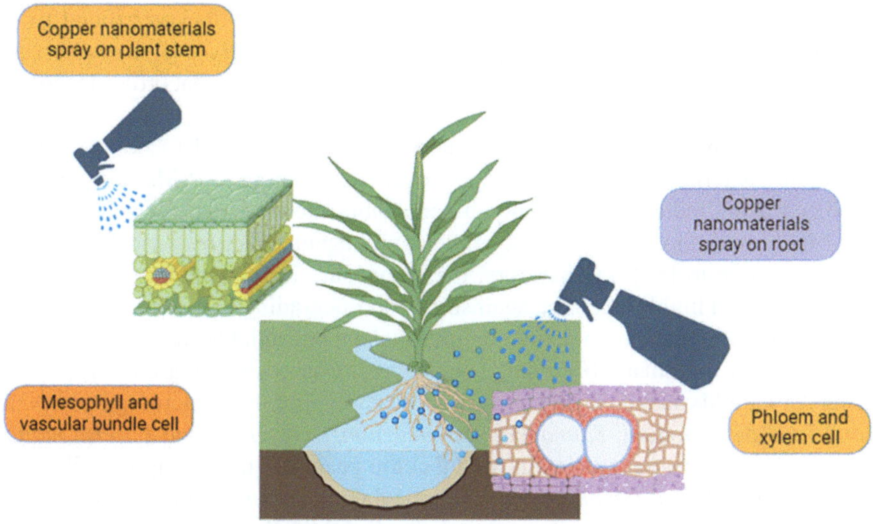

Fig. 1 Copper nanomaterials increase plant growth

utilized in the food industry, was employed as a stabilizing agent. We know the two substances have not been combined to create water-dispersible CuNMs. On the other hand, ascorbic acid alone was utilized as the reducing agent in other instances [36] while citrate and poly (vinyl alcohol) were utilized as stabilizing agents [71]. Nevertheless, the described powders showed signs of significant agglomeration or micro-sized particles, making them unsuitable for fungicide use (Fig. 1).

3 Copper-Based Nanomaterials for Controlling Pathogens

According to research by Wang et al. [69], CuNMs exhibit broad-spectrum antibacterial activity, and the incorporation of a hybrid composite consisting of cellulose and CuNMs into polyvinyl alcohol (PVA) film enhances the antibacterial efficacy of these NMs. This advancement highlights the potential for new materials with increased antibacterial properties, emphasizing the synergistic effects of these hybrid composites in diverse applications and CuNMs and cellulose function as nanofillers in this instance [73]. According to [32], food deterioration-causing microbes such as *Staphylococcus aureus* and *Escherichia coli* are susceptible to the antibacterial properties of the copper-coated cellulose film. Jia et al. [32] reported more antibacterial activity against S. aureus than *E. coli.* A different study determined whether chitosan-coated CuNMs might suppress tomato-harming fungi such as *Fusarium oxysporum* and *Alternaria solani* [67]. The antibacterial activity of chemically synthesized CuNMs was reported by [59] against fungi and bacteria, including *Aspergillus niger*, *Aspergillus flavus*, *Candida albicans*, and *Micrococcus luteus*, *S. aureus*, *E. coli*,

Klebsiella pneumoniae, and *Pseudomonas aeruginosa*. CuNMs were found to be quite efficient against the following bacteria: *E. coli*, *C. albicans*, *S. aureus*, *M. luteus*, *A. niger*, *K. pneumoniae*, *A. flavus*, and *P. aeruginosa*, in that order, according to the investigators [59].

According to [45], *Bacillus cereus* and *E. coli* were found to be susceptible to the antibacterial effects of a composite film containing CuNMs and chitosan stabilized using iodine. Additionally, a nanocomposite film made of cellulose and CuNMs showed remarkable antibacterial activity against *S. aureus* and *K. pneumoniae*, according to [53]. Furthermore, the composite film, including cellulose and CuNMs, proved highly effective against the food-degrading bacteria *Pseudomonas sp.*, these films-based antibacterial qualities make them suitable for food packaging applications [42]. Remarkably, a nanocomposite film based on guar gum and incorporating Ag-CuNMs has shown antibacterial efficacy against three microorganisms that cause food deterioration: *Salmonella typhimurium*, *S. enterica*, and *Listeria monocytogenes*. Furthermore, studies showed that the guar gum nanocomposite film was more effective against *S. typhimurium* in particular than it was against *S. enterica* and *L. monocytogenes*. [2]. Some ideas suggest that CuNMs' direct contact with bacterial cells increases their ability to inhibit bacterial growth (because CuNMs and bacteria transfer electrons more readily). In this instance, the metallic NMs and the slightly negative bacteria both function as electron acceptors, facilitating the transfer of electrons and the bacterial membrane's rupture. A photocatalytic mechanism may also be responsible for killing the bacteria, as evidenced by the excited electron–hole pairs that can result from light irradiation in copper oxide (CuO). Furthermore, copper's antibacterial properties come from cellular damage that occurs when liberated Cu^{2+} ions touch bacterial membranes. CuNMs, however, are more prone to oxidation and, therefore, extra unstable. CuO NMs have higher stability but somewhat lower activity; to solve this issue, CuNMs were changed into CuO NMs [40].

4 CuNMs-Based Antibacterial Activity

As already noted, research has been done on the antibacterial properties of CuNMs, but their mechanisms have not widely understood. Some papers have described the mechanism of the action for CuNMs as antibacterial agents. Carboxyl group present on microbial surface attracts CuNMs to them, enabling them to bind to the bacterial cell membrane (Ingel et al. 2014; [57]). Lemire et al. [41] and Swarankar et al. [66] demonstrated that different factors contribute to the antibacterial activities of NMs, such as the generation of reactive oxygen species (ROS), changes in membranes, enzyme inhibition, and dysfunctioning proteins [54]. Cellular metabolic pathways destruction and production of pits within the oxidative stress of the membrane have resulted in the death of cells by interacting with Cumns, as reported by Deryabin et al. [18] and Shende et al. [62]. An alternative form of an antibacterial agent was suggested involving Cu polymer nanocomposites. Scientists showed that nanocomposites possessed bactericidal activity in removing copper ions and CuNMs into

Fig. 2 Copper nanomaterials reduce fungal disease and increase plant growth

media after synthesis. Such copper ions can interact with amines-carboxyl groups-sulfhydril groups in the murein layer upon their outside contact with released Cu ions, leading to protein denaturation. Similarly, the Cu^{2+} ions can also bind to DNA and thus promote cross linking of DNA strands, which disrupt their spiral structure. In addition, copper NMs induce oxidative stress, produce membrane pits, interfere with cell metabolic pathways, and finally cause apoptosis [18, 62]. The researchers stated that Cu polymer nanocomposites were considered a promising antibacterial drug candidate. They argued that removing copper ions and CuNMs from these nanocomposites makes them bactericidal in nature. According to their research, Cu ions remove and CuNMs from nanocomposites is what gives them their bactericidal action. The protein is denatured due to the liberated Cu ions' interactions with amines, carboxyl groups, and sulfhydryl groups in the bacterial cell wall layer when they come into contact with an outer bacterial membrane. The helical structure becomes disorganized due to Cu ions (Cu^{2+}) binding to DNA and intercalating nucleic acid strands (Fig. 2).

5 Cu-Based NMs for Controlling Insect Pests and Diseases

Cu is well-known for its usage in agriculture as a plant defender against bacterial and fungal infections. As previously mentioned, a range of plant illnesses caused by bacteria, fungi, viruses, and insects are thought to cause a third of the world's annual plant harvest disappearance. Cu has been used in medicine since ancient times to fight

Table 1 Different microbial strain causes disease

Microbial strain	Disease	Host	References
Phytopthora infestans	Potato late blight fungus	Solanum lycopersicum	Giannousi et al. [23]
Xanthomonas axonopodispv. Punicae	Bacterial blight	Pomegranate	Mondal et al. [46]
B. cinerea	Gray mold	Tomato	Malandrakis et al. [44]
M. aruticola	Brown rot	Peach	Malandrakis et al. [44]
A. alternata	Black rot	Tomato	Malandrakis et al. [44]

against the threat of illness. One such substance is copper sulfate, which has antifungal qualities and is a main component of most fungicides for gardens and farms sold commercially. Typically, the plants were sprayed with a solution of copper sulfate, lime, or soda ash. According to [9], Cu-based fungicides are essential for preventing disease in a wide range of plant species because they produce hydroxyl radicals that are highly reactive and that can harm proteins, lipids, DNA, and other macromolecules. Moreover, CuO serves as a protective agent for crucial plant species like tea, bananas, cocoa, citrus, and coffee, shielding them against prominent diseases of fruit and fungal leaf such as blight, downy or mildew of powdery, as well as corrosion [37]. Cu serves as a plant protector in sustainable agriculture as well. Over the past 15–20 years, the organic farming industry has made significant efforts to reduce the amount of applied Cu and seek a copper substitute for crop protection. But now, only so many active ingredients, strategies, or approaches could replace Cu. Despite the challenges associated with investigating Cu alternatives, more studies must be done in the next few years to develop better techniques and agronomic strategies that minimize Cu [70]. The efficient utilization of advanced technologies like liposomes, nanoemulsions, and micro-emulsions in agrochemical formulations has significantly reduced the dependence on petrochemicals for delivering fungicides, herbicides, and insecticides [12]. Consequently, the application of nanotechnology to plant protection has been increasingly prominent in recent years [16]. Additionally, Cu nanoparticle (CuNPs) containing soda lime glass powder with a low melting point showed significant antibacterial action against fungus and bacteria [21] (Table 1).

6 Cu-Based NMs for Controlling *Phytopthora Infestans*

According to the studies, biocompatible surfactants such as polyoxyethylene, sorbitan laurate, and polyethylene glycol were present when the copper nanoparticles were created for specific requirements and can be used with low production and active ingredients and are more effective than commercial agricultural chemicals. Additionally, phytotoxicity tests show that it does not harm plants, especially when tested in the fields of tomato and potato [23].

7 Cu-Based NMs for Controlling *Xanthomonas Axonopodispv*

Xanthomonas axonopodispv causes pomegranate (*Punica granatum*) wilt disease. In India, Punicae (Xap) is a significant pomegranate disease. Here, we discovered that copper nanomaterials may effectively stop the growth of bacteria that are carried by water, thanks to Xap. At a concentration of 0.2 ppm, nano-copper inhibits Xap development, which is 10,000 times less than what copper oxychloride is advised for. Scanning electron microscope showed that the cell walls of nano-copper-treated bacteria were disturbed, unable to form tissue and had strong water immersion properties. This study shows that CuNMS may be an environmentally friendly strategy to control pomegranate blight [46].

8 Cu-Based NMs for Controlling the Plant Pathogens

In vitro poison agar experiments were used to determine the mean inhibitory doses of Cu, CuO, silver (Ag), and zinc oxide (ZnO) NMs and assess the efficacy of NMs against several significant plant-pathogenic fungi. Plant parasite-related crop losses represent a significant global problem to the current agricultural production system, with illnesses responsible for at least 25% of the losses. Mycelial growth inhibition experiments conducted in vitro showed notable differences in sensitivity between the NMs examined and fungal strains from various genera. The sequence of NMs' toxicity to fungal strains was CuNMs > ZnO-NMs > Ag NMs > CuO NMs [44].

The increased antibacterial action is attributed to the disruptive complementary effect of calcium dioxide leached from the glass. In an equivalent study evaluating responses against fungi on the tomato infected with *Phytophthora infestans*, Cu-based NMs displayed superior activity over commercial agrochemicals at lower concentrations [23]; similarly, in another investigation targeting *Xanthomonas axonopodispv*. Punicae, the pathogen causing bacterial blight in pomegranates, nano-copper effectively countered the bacteria at a concentration nearly 10, thousand times lower than the recommended amount for Cu-oxychloride [46]. Cu has a synergistic effect with chitosan complex nano-gels, which inhibits the growth of *Fusarium graminearum*, a plant disease that affects cereal plants. As [11] have argued, these nanohydrogels are biocompatible, making them a distinctive category of new Cu-based biopesticides. In vitro antifungal activity of chemically size synthesized CuNMs against four plant pathogens, namely *Phoma destructiva, Alternaria alternata, F.*

oxysporum, and *Curvularia lunata*, was evaluated by [34]. A study that found significant antifungal activities on the test fungi [62] proposed incorporating CuNMs into nanofibrate for use as antifungal agents. Moreover, inhibition of *Fusarium graminearum*, *Fusarium oxysporum*, and *Fusarium culmorum* growth was observed by biogenic synthesized CuNMs [62]. Furthermore, suppression in the growth of other fungal strains, including *F. equiseti*, *F. oxysporum*, and *F. culmorum*, has also been observed in chemically synthesized CUNMs [10]. Although several nano-based products are still under development, some, like nano-pesticides, nano-fungicides, and nano-insecticides, are already available in the market [15]. CuNMs will, therefore, be the most sought-after nano-formulations in agriculture, and they may be applied in various ways to plant protection in forthcoming years.

9 Conclusions and Future Perspectives

Several researches supported that CuNMs offer improved plant protection as well as use of CuNPs is more effective than use of bulk copper in treating a range of common bacterial and fungal diseases. CuNMS cause pathogens to die by breaking down their cell walls and changing cell organelles' operations. As a result, copper at the nanoscale has advantages over other materials when it comes to treating plant illnesses. In addition, food packaging can also benefit from its application to prolong food items of the shelf life. A sustainable method involving the chemical Cu ions reduced with the antioxidant vitamin C, complemented by sodium citrate as a protective agent, was effectively employed for CuNPs synthesis. These green-synthesized CuNMs exhibited potent antifungal properties against *F. oxysporum*, *F. solani*, and *Neofusicoccum sp.* A number of investigations have asserted that when CuNMS are utilized more than a specific concentration, they likewise exhibit detrimental impacts on plants. Applying CuNMs at a higher concentration may impact plant metabolism and yield. Therefore, it is crucial to thoroughly understand the relationship between nanomaterials and plants to employ CuNMs in agriculture in a safer manner. CuNMs hold significant promise with the ability to increase food shelf life and decrease waste in agriculture and the food sector, which is crucial for meeting the growing demands of the global population. However, ensuring their effective and responsible utilization is key to realizing these benefits.

10 Competing Interests

The authors declare that they have no competing interests.

11 Declaration of Generative AI and AI-Assisted Technologies in the Writing Process

During the preparation of this work the book chapter author(s) have not taken any assistance from AI, and the book chapter author(s) reviewed and edited the content as needed and take(s) full responsibility for the content of the publication. All authors read and approved the final version of the manuscript.

References

1. Acharya A, Pal PK (2020) Agriculture nanotechnology: translating research outcome to field applications by influencing environmental sustainability. NanoImpact 19:100232
2. Arfat YA, Ejaz M, Jacob H, Ahmed J (2017) Deciphering the potential of guar gum/Ag-Cu nanocomposite films as an active food packaging material. Carbohydr Polym 10(157):65–71
3. Bachheti A, Bachheti RK, Husen A (eds) (2024) Carbon-based nanomaterials: synthesis, agricultural, biomedical, and environmental interventions. Springer Nature Singapore. https://doi.org/10.1007/978-981-97-0240-4
4. Bachheti RK, Bachheti A, Husen A (eds) (2024a) Metal and metal-oxide based nanomaterials: Singapore: synthesis, agricultural, biomedical and environmental interventions. Springer Nature Singapore. https://doi.org/10.1007/978-981-99-7673-7
5. Bachheti RK, Bachheti A (eds) (2023) Secondary metabolites from medicinal plants: nanoparticles synthesis and their applications. CRC Press, Boca Raton, USA. https://doi.org/10.1201/9781003213727
6. Bachheti RK, Bachheti A, Husen A (eds) (2023a) Nanomaterials for environmental and agricultural sectors. Springer, Singapore. https://doi.org/10.1007/978-981-99-2874-3
7. Banik S, Pérez-de-Luque A (2017) In vitro effects of copper nanomaterials on plant pathogens, beneficial microbes and crop plants. Span J Agric Res15(2):1005
8. Baker S, Volova T, Prudnikova SV, Satish S, Prasad N (2017) Nanoagroparticles emergingtrends and future prospects in the modern agriculture system. Environ Toxicol Pharmacol 53:10–17
9. Borkow G, Gabbay J (2005) Copper as a biocidal tool. Curr Med Chem 12:2163–2175
10. Bramhanwade K, Shende S, Bonde S, Gade A, Rai M (2016) Fungicidal activity of Cu nanomaterials against fusarium causing crop diseases. Environ Chem Lett 14:229–235
11. Brunel F, El Gueddari NE, Moerschbacher BM (2013) Complexation of copper (II) with chitosan nanogels: towards control of microbial growth. Carbohydr Polym 92:1348–1356
12. Castro ML, Ojeda C, Cirelli A (2014) Advances in surfactants for agrochemicals. Environ Chem Lett 12:85–95
13. Chatterjee AK, Chakraborty R, Basu T (2014) Mechanism of antibacterial activity of copper nanomaterials. Nanotechnology 25(13):135101
14. Chen J, Mao S, Xu Z, Ding W (2019) Various antibacterial mechanisms of Bio-synthesized copper oxide nanomaterials against soilborne *Ralstonia solanacearum*. RSC adv 9(7):3788–3799
15. Chinnamuthu CR, Boopathi PM (2009) Nanotechnology and agroecosystem. Madras Agri J 96:17–31
16. Chowdappa P, Gowda S (2013) Nanotechnology in crop protection: status and scope. Pest Manage Hort Ecosyst 19:131–151
17. Dasgupta N, Ranjan S, Mundekkad D, Ramalingam C, Shanker R, Kumar A (2015) Nanotechnology in agro-food: from field to plate. Food Res Int 69:381–400
18. Deryabin DG, Aleshina ES, Vasilchenko AS, Deryabin TD, Efremova LV, Karimov IF, Korolevskay LB (2013) Investigation of copper nanomaterials antibacterial mechanisms tested by luminescent *Escherichia coli* strains. Nanotechnol Russia 8:402–408

19. Elmer WH, White JC (2016) The use of metallic oxide nanomaterials to enhance growth of tomatoes and eggplants in disease infested soil or soilless medium. Environ Sci Nano 3(5):1072–1079
20. Elmer W, White JC (2018) The future of nanotechnology in plant pathology. Annu Rev Phytopathol 56:111–133
21. Esteban-Tejeda L, Malpartida F, Esteban-Cubillo A, Pecharroman C, Moya JS (2009) Antibacterial and antifungal activity of a soda-lime glass containing copper nanomaterials. Nanotechnology 20:505701
22. Gawande MB, Goswami A, Felpin FX, Asefa T, Huang X, Silva R, Zou X, Zboril R, Varma RS (2016) Cu and Cu-based nanomaterials: synthesis and applications in catalysis. Chem Rev 116(6):3722–3811
23. Giannousi K, Avramidis I, Dendrinou-Samara C (2013) Synthesis, characterization and evaluation of copper based nanomaterials as agrochemicals against *Phytophthora infestans*. RSC Adv 3:21743–21752
24. Gogos A, Knauer K, Bucheli TD (2012) Nanomaterials in plant protection and fertilization: current state, foreseen applications, and research priorities. J Agric Food Chem 60(39):9781–9792
25. Gomez A, Narayan M, Zhao L, Jia X, Bernal RA, Lopez-Moreno ML, Peralta-Videa JR (2021) Effects of nano-enabled agricultural strategies on food quality: current knowledge and future research needs. Mater 401:123385
26. Husen A, Bachheti RK, Bachheti A (eds) (2023) Current trends in green nano-emulsions: food, agriculture and biomedical sectors. Springer Nature Singapore, Singapore. https://doi.org/10.1007/978-981-99-5398-1
27. Husen A (2022) Engineered nanomaterials for sustainable agricultural production, soil improvement and stress management. Elsevier Inc., 50 Hampshire St., 5th Floor, Cambridge, MA 02139, USA. https://doi.org/10.1016/C2021-0-00054-7
28. Husen A (2023) Nanomaterials and nanocomposites exposures to plants (Response, interaction, phytotoxicity and defense mechanisms). Springer Nature Singapore Pte Ltd., 152 Beach Road, #21-01/04 Gateway East, Singapore 189721, Singapore. https://doi.org/10.1007/978-981-99-2419-6
29. Husen A (2024) Plant response to gold nanoparticles. Springer Nature Singapore Pte Ltd., 152 Beach Road, #21-01/04 Gateway East, Singapore 189721, Singapore
30. Husen A (2025a) Plant response to silver nanoparticles. Springer Nature Singapore Pte Ltd., 152 Beach Road, #21-01/04 Gateway East, Singapore 189721, Singapore
31. Husen A (2025b) Emerging carbon nanomaterials for sustainable agricultural practices (Synthesis, plant growth, performance, production and protection). Springer Nature Singapore Pte Ltd., 152 Beach Road, #21-01/04 Gateway East, Singapore 189721, Singapore
32. Jia B, Mei Y, Cheng L, Zhou J, Zhang L (2012) Preparation of copper nanomaterials coated cellulose films with antibacterial properties through one-step reduction. ACS Appl Mater Interfaces 6:2897–2902
33. Jin-Chul K, Madhusudhan A, Husen A (2021) Smart nanomaterials in biomedical applications. Springer Nature Switzerland AG, Gewerbestrasse 11, 6330 Cham, Switzerland. https://doi.org/10.1007/978-3-030-84262-8
34. Kanhed P, Birla S, Gaikwad S, Gade A, Seabra AB, Rubilar O, Duran N, Rai M (2014) In vitro antifungal efficacy of copper nanomaterials against selected crop pathogenic fungi. Mater Lett 115:13–17
35. Keller AA, Adeleye AS, Conway JR, Garner KL, Zhao L, Cherr GN, Hong J, Gardea-Torresdey JL, Godwin HA, Hanna S, Ji Z (2017) Comparative environmental fate and toxicity of copper nanomaterials. NanoImpact 7:28–40
36. Khanna PK, Gaikwad S, Adhyapak PV, Singh N, Marimuthu R (2007) Synthesis and characterization of copper nanomaterials. Mater Lett 61(25):4711–4714
37. Kiaune L, Singhasemanon N (2011) Pesticidal copper (I) oxide: environmental fate and aquatic toxicity. Rev Environ Contam Toxicol 213:1–26

38. Kim SW, Kim KS, Lamsal K, Kim YJ, Kim SB, Jung MY, Sim SJ, Kim HS, Chang SJ, Kim JK, Lee YS (2009) An in vitro study of the antifungal effect of silver nanomaterials on oak wilt pathogen *Raffaelea* sp. J Microbiol Biotechnol 19(8):760–764
39. Khot LR, Sankaran S, Maja JM, Ehsani R, Schuster EW (2012) Applications of nanomaterials in agricultural production and crop protection: a review. Crop Prot 35:64–70
40. Lara HH, Ayala-Nuñez NV, Ixtepan-Turrent L, Rodriguez-Padilla C (2010) Mode of antiviral action of silver Nanomaterials against HIV-1. J Nanobiotechnol 8(1):1
41. Lemire JA, Harrison JJ, Turner RJ (2013) Antimicrobial activity of metals: mechanisms, molecular targets and applications. Nat Rev Microbiol 11(6):371–384
42. Longano D, Ditaranto N, Cioffi N, Di Niso F, Sibillano T, Ancona A, Conte A, Del Nobile MA, Sabbatini L, Torsi L (2012) Analytical characterization of laser-generated copper nanomaterials for antibacterial composite packaging food. Anal Bioanal Chem 403:1179–1186
43. Mahmoodi S, Elmi A, Hallaj-Nezhadi S (2018) Copper nanomaterials as antibacterial agents. J Mol Pharm Org Process Res 6(1):1–7
44. Malandrakis AA, Kavroulakis N, Chrysikopoulos CV (2019) Use of copper, silver and zinc nanomaterials against foliar and soil-borne plant pathogens. Sci Total Environ 670:292–299
45. Mallick S, Sharma S, Banerjee M, Ghosh SS, Chattopadhyay A, Paul A (2012) Iodine-stabilized Cu nanoparticle chitosan composite for antibacterial applications. ACS Appl Mater Interfaces 4(3):1313–1323
46. Mondal K, Mani C (2012) Investigation of the antibacterial properties of nano-copper against *Xanthomonas axonopodis* pv. punicae, the incident of pomegranate bacterial blight. Ann Microbiol 62:889–893
47. Madhusudhan A, Purohit SD, Prasad R, Husen A (2024) Functional smart nanomaterials and their theranostics approaches. Springer Nature Singapore Pte Ltd., 152 Beach Road, #21-01/04 Gateway East, Singapore 189721, Singapore
48. Mandal AK, Ghorai S, Husen A (2023) Functionalized smart nanomaterials for point-of-care testing. Springer Nature Singapore Pte Ltd., 152 Beach Road, #21-01/04 Gateway East, Singapore 189721, Singapore
49. Nunes I, Jacquiod S, Brejnrod A, Holm PE, Johansen A, Brandt KK, Priemé A, Sørensen SJ (2016) Coping with copper: legacy effect of copper on potential activity of soil bacteria following a century of exposure. FEMS Microbiol Ecol 92(11):175
50. Park JW, Lee IC, Shin NR, Jeon CM, Kwon OK, Ko JW, Kim JC, Oh SR, Shin IS, Ahn KS (2016) Copper oxide nanomaterials aggravate airway inflammation and mucus production in asthmatic mice via MAPK signaling. Nanotoxicology 10(4):445–452
51. Pariona N, Mtz-Enriquez AI, Sánchez-Rangel D, Carrión G, Paraguay-Delgado F, Rosas-Saito G (2019) Green-synthesized copper nanomaterials as a potential antifungal against plant pathogens. RSC Adv 9(33):18835–18843
52. Parveen F, Sannakki B, Mandke MV, Pathan HM (2016) Copper Nanomaterials: synthesis methods and its light harvesting performance. Sol Energy Mater Sol Cells 144:371–382
53. Pinto RJ, Daina S, Sadocco P, Neto CP, Trindade T (2013) Antibacterial activity of nanocomposites of copper and cellulose. Bio Med res Int
54. Raffi M, Mehrwan S, Bhatti TM, Akhter JI, Hameed A, Yawar W, ul Hasan MM (2010) Investigations into the antibacterial behavior of copper Nanomaterials against *Escherichia coli*. Ann Microbiol 60:75–80
55. Rajpal VR, Dhingra Y, Khungar L, Mehta S, Minkina T, Rajput VD, Husen A (2024) Metal and metal-oxide nanoparticles in nanosensing as well as managing biotic stress in plant system. Current Res Biotechnol 7:100219. https://doi.org/10.1016/j.crbiot.2024.100219
56. Rath SK, Dwivedi V, Husen A, Akhtar N (2024) Smart nanomaterials for drug delivery and neurological diseases management. Springer Nature Singapore Pte Ltd., 152 Beach Road, #21-01/04 Gateway East, Singapore 189721, Singapore ISBN: 978-981-97-0307-4. https://doi.org/10.1007/978-981-97-0308-1
57. Ren G, Hu D, Cheng EW, Vargas-Reus MA, Reip P, Allaker RP (2009) Characterisation of copper oxide nanomaterials for antimicrobial applications. Int J Antimicrob agents 33(6):587–590

58. Rai M, Ingle AP, Pandit R, Paralikar P, Shende S, Gupta I, Biswas JK, da Silva SS (2018) Copper and copper nanomaterials: role in management of insect-pests and pathogenic microbes. Nanotechnol Rev 7(4):303–315
59. Ramyadevi J, Jeyasubramanian K, Marikani A, Rajakumar G, Rahuman AA (2012) Synthesis and antimicrobial activity of copper nanomaterials. Mater Lett 71:114–165
60. Savary S, Ficke A, Aubertot JN, Hollier C (2012) Crop losses due to diseases and their implications for global food production losses and food security. Food Sec 4:519–537
61. Servin A, Elmer W, Mukherjee A, De la Torre-Roche R, Hamdi H, White JC, Bindraban P, Dimkpa C (2015) A review of the use of engineered nanomaterials to suppress plant disease and enhance crop yield. J Nanopart Res 17:1–21
62. Shende S, Ingle AP, Gade A, Rai M (2015) Green synthesis of copper Nanomaterials by *Citrus medica* Linn. (Idilimbu) juice and its antimicrobial activity. World J Microbiol Biotechnol 3:865–873
63. Singh DP, Ojha AK, Srivastava ON (2009) Synthesis of different Cu $(OH)_2$ and CuO (nanowires, rectangles, seed-, belt-, and sheetlike) nanostructures by simple wet chemical route. J Phys Chem 113(9):3409–3418
64. Skandalis N, Dimopoulou A, Georgopoulou A, Gallios N, Papadopoulos D, Tsipas D, Theologidis I, Michailidis N, Chatzinikolaidou M (2017) The effect of silver nanomaterials size, produced using plant extract from *Arbutus unedo*, on their antibacterial efficacy. Nanomaterials 7(7):178
65. Stadler T, Buteler M, Weaver DK, Sofie S (2012) Comparative toxicity of nanostructured alumina and a commercial inert dust for *Sitophilus oryzae* (L.) and *Rhyzopertha dominica* (F.) at varying ambient humidity levels. J Stored Prod Res 48:81–90
66. Swarnkar RK, Pandey JK, Soumya KK, Dwivedi P, Sundaram S, Prasad S, Gopal R (2016) Enhanced antibacterial activity of copper/copperoxide nanowires prepared by pulsed laser ablation in water medium. Appl Phys A 122:1–7
67. Saharan V, Sharma G, Yadav M, Choudhary MK, Sharma SS, Pal A, Raliya R, Biswas P (2015) Synthesis and in vitro antifungal efficacy of Cu–chitosan nanomaterials against pathogenic fungi of tomato. Int J Biol lmacromol 75:346–353
68. Vitanovic E (2012) Use of Cu fungicides in vineyards and olive groves. INTECH Open Access Publisher. Wang L, Hu C, Shao L (2017) The antimicrobial activity of nanomaterials: present situation and prospects for the future. Int J Nanomedicine 1227–1249
69. Wang P, Lombi E, Zhao FJ, Kopittke PM (2016) Nanotechnology: a new opportunity in plant sciences. Trendsplant Sci 21(8):699–712
70. Wilbois KP, Kauer R, Fader B, Kienzle J, Haug P, Fritzsche-Martin A, Drescher N, Reiners E, Rohrig F (2009) Copper as a plant protection product with special regards to organic farming. J Kulturpflanzen 61:140–152
71. Wu S (2007) Preparation of fine copper powder using ascorbic acid as a reducing agent and Its application in MLCC. Mater Lett 61(4–5):1125–1129
72. Yadav L, Tripathi RM, Prasad R, Pudake RN, Mittal J (2017) Antibacterial activity of Cu nanomaterials against *E. coli*, *Staphylococcus aureus* and *Pseudomonas aeruginosa.* Nano Biomed Eng 9(1):9–14
73. Zhong T, Oporto GS, Jaczynski J, Jiang C (2015) Nanofibrillated cellulose and copper nanomaterials embedded in polyvinyl alcohol films for antimicrobial applications. Bio Med Res Int

Use of Copper-Based Nanomaterials in Postharvest Management

Prakash Chandra Gupta, Monika Dwivedi, Vishal Chand, Nisha Sharma, Priyanka Mishra, and Shivam Verma

Abstract The management of postharvest diseases in fruits and vegetables is imperative due to the significant value-added nature of these commodities and the substantial economic losses associated with spoilage. The utilization of synthetic fungicides is widely preferred on a global scale for the management of postharvest illnesses affecting fruits and vegetables. Nevertheless, these methods have been hindered by many issues and limitations, prompting researchers to devise new strategies for managing postharvest infections. Recently, nanomaterial treatments have shown encouraging outcomes, and they are being explored for their ability to decrease the use of synthetic fungicides for controlling postharvest rot in fruits and vegetables. Using copper-based nanomaterials in postharvest management signifies a pivotal shift in sustainable farming practices, epitomizing the transformative force of nanotechnology in agriculture. This paradigm shift, supported by ongoing research and technological advancements, holds promise for addressing global challenges in food security, resource efficiency, and environmental sustainability. In postharvest management, nanotechnology has emerged as a strategic tool for alleviating losses and optimizing supply chain efficiency. Realizing the full benefits of nanotechnology in agriculture requires sustained interdisciplinary collaboration and a responsible approach to implementation, ensuring both innovation and risk mitigation. The comprehensive content included in this chapter encompasses copper-based nanomaterials employed to manage postharvest decay.

Keywords Nanomaterial · Copper · Nanotechnology · Harvest management · Fungicides

P. C. Gupta (✉) · N. Sharma · P. Mishra · S. Verma
School of Pharmaceutical Sciences, Chhatrapati Shahu Ji Maharaj University, Kanpur, Uttar Pradesh, India
e-mail: prakashchandragupta@csjmu.ac.in

M. Dwivedi
Department of Pharmaceutical Sciences & Technology, Birla Institute of Technology, Mesra, Ranchi, Jharkhand, India

V. Chand
School of Life Sciences and Biotechnology, Chhatrapati Shahu Ji Maharaj University, Kanpur, Uttar Pradesh, India

R. K. Bachheti et al. (eds.), *Copper-Based Nanomaterial for Agricultural Practices*, Smart Nanomaterials Technology, https://doi.org/10.1007/978-981-96-8610-0_10

1 Introduction

Fruit and vegetable postharvest infections are frequently attributed to various plant pathogenic microorganisms, such as fungi and bacteria, resulting in significant losses throughout storage and transportation. Nanotechnology, a novel methodology, has been employed in various scientific domains, including agriculture. Nanomaterials provide remarkable characteristics that render them a viable option for sustainable horticulture, particularly in addressing postharvest illnesses affecting fruits and vegetables, both at the scientific and industrial levels. Copper plays a critical role as a micronutrient and is essential for sustaining the health and nutrition of plants. Copper nanoparticles (Cu-NPs) possess cost-effective attributes, unique nanoscale characteristics, established antibacterial activity, and wide-range applicability. These qualities make them highly promising for agricultural activities, particularly pest management and disease control. Both copper nanoparticles and copper oxide nanoparticles (CuO-NPs), which are copper-based nanoparticles, are being increasingly incorporated into nano-formulations and agrochemicals. This integration facilitates the efficient delivery and controlled release of pesticides and fertilizers. The applications of copper nanoparticles include plant protection, the stimulation of plant growth, their use as nanosensors, and their effective use as antifungal and antibacterial agents [26].

This chapter extensively explores the diverse applications of copper nanoparticles, focusing on agricultural progress and the long-term storage of agricultural products. It has investigated their use in agrochemical formulations to prevent plant diseases and their integration into nanosensors designed for detecting pesticides, chemicals, and toxins (Fig. 1).

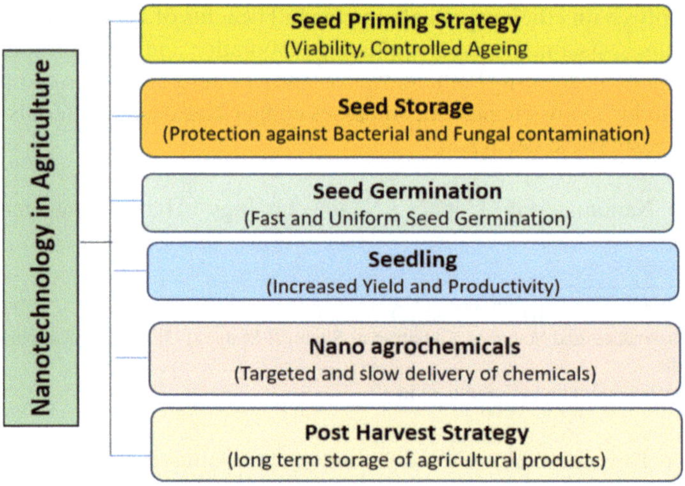

Fig. 1 Nanotechnology in cooperation with different steps of agriculture

In a related context, using copper-based nanomaterials in postharvest management represents a pivotal shift in sustainable farming practices, showcasing the transformative impact of nanotechnology in agriculture. The integration of nanomaterials at the nanoscale introduces unique properties that, when applied in agriculture, can enhance critical aspects such as crop yield, nutrient absorption, and disease resistance. In the context of postharvest management, nanotechnology has emerged as a strategic tool for alleviating losses and optimizing supply chain efficiency.

Postharvest management plays a pivotal role in ensuring the availability of safe and high-quality food for an ever-growing global population. Despite advancements in agricultural practices, postharvest losses remain a significant challenge, threatening food security and economic sustainability. This chapter explores the potential of copper-based nanomaterials to address these challenges, leveraging their unique properties to revolutionize postharvest management practices.

1.1 Postharvest Losses and Challenges

Postharvest loss is a significant factor in exacerbating food scarcity, with approximately 50% of the annual harvest succumbing to losses across the harvesting phases, postharvest handling, storage, processing, distribution, and consumption. It is essential to employ appropriate postharvest packaging, transportation, and storage practices to mitigate the impact of postharvest loss. Challenges or obstacles, including insufficient storage infrastructure, suboptimal transportation systems, and susceptibility to microbial contamination, constitute pivotal factors contributing to the detriment of postharvest preservation, particularly in the context of nanoparticles consisting of copper. The resolution of these challenges is paramount for ensuring the sustenance of food supplies while concurrently mitigating the associated economic and environmental repercussions.

1.2 Nanotechnology in Agriculture

The evolution of technologies involving nano-sized materials, with dimensions of 10^{-9} m, has progressed remarkably from initial interest to diverse applications. The main impetus behind the rapid advancement in this field stems from the unique phenomenon exhibited by nanomaterials, opening avenues for novel applications across industries such as electronics, health, food, and agriculture [38]. Owing to their improved surface area-to-volume ratio, elevated surface energy, and quantum confinement, they exhibit unique optical, electronic, chemical, mechanical, and magnetic properties compared to those of their bulk equivalents [77]. Among the various types of nanoparticles, metallic nanoparticles (MNPs are notably fascinating, offering a wide range of applications across various fields. They are utilized in the food industry [45, 69, 74], medical diagnostics [7, 21], antimicrobial treatments [10,

82, 83], health therapies [89], antioxidants [75], wastewater treatment [34], heavy-metal detection [48], environmental cleanup [67], and agricultural production [35, 42], and crop protection [36, 41]. The integration of nanotechnology in agriculture, often referred to as nanotechnology in agriculture (Agri-Nano), holds immense potential to revolutionize farming practices, enhance crop yields, and mitigate environmental challenges. Enhancing global food security, mitigating malnutrition, and minimizing the ecological impact of agriculture necessitate substantial improvements in our food systems. In this context, agro-nanotechnology has emerged as a promising avenue for fostering sustainable agriculture and advancing crop development. The integration of nanotechnology into the agro-food industry holds significant potential through the introduction of innovative active ingredients. Notable examples include the utilization of nano-fertilizers and micronutrients, the incorporation of nano-carrier systems enveloping agrochemicals, and the deployment of nano-pesticides, which encompass environment-friendly options such as green and biopesticides. The prevalent use of conventional agrochemicals is associated with substantial adverse environmental consequences, posing threats to both terrestrial and aquatic ecosystems due to their widespread dispersion and leaching. Addressing these challenges involves strategically implementing controlled release mechanisms for active components in agrochemicals. This strategy aims to regulate the quantity and frequency of applications in the field, thereby mitigating issues associated with the excessive use of agrochemicals. In contrast to traditional agrochemicals, the efficacy of copper-based nanoparticles warrants comprehensive evaluation to address inherent challenges within the agricultural sector. Their attractiveness lies in their affordability compared to widely studied metals such as silver, gold, palladium, and platinum, which makes them easily scalable. Additionally, copper-based nanoparticles provide more significant physical and chemical stability than silver and can be effortlessly integrated with polymers [53]. The biotransformation of *Pseudomonas fluorescens* MAL2, which closely resembles *Pseudomonas fluorescens* DSM 12442 T DSM, has proven to be highly effective in combating *T. castaneum*, as demonstrated by [24]. Another noteworthy example involves the synthesis of CuO-NPs through the utilization of metabolites produced by the *Aspergillus niger* strain (G3-1). This method resulted in mortality percentages ranging from 55% to 94.4% for *S. granaries* and 70–90% for *R. dominica*, showing its potential efficacy [8].

Copper-based nanoparticles possess distinct properties, enabling their use in various fields such as electronics [44], optical devices [47], nanocomposite coatings [3], thermal transfer nanofluids [23], catalysis (Gawande et al., 2016), and pollutant degradation [68], and most probably in agricultural advancement to mitigate all the shortcomings linked with traditional agricultural methodologies. Cu-based nanogenic agrochemicals exhibit potential applications in promoting plant growth and managing agricultural diseases, presenting a more sustainable alternative to their conventional counterparts.

1.3 Copper Oxide Nanomaterials

Copper oxide nanoparticles, including cuprous oxide (Cu_2O) and cupric oxide (CuO), represent another class of copper-based nanomaterials. These compounds exhibit distinct electronic and optical properties, rendering them suitable for applications in catalysis, sensors, and electronic devices [40, 81]. CuO-NPs were synthesized using sodium hydroxide and copper nitrate, and characterized using XRD, SEM, and UV–visible spectroscopy. The particles, with an average size of 23 nm, enhanced growth parameters, including root length, shoot length, and plant biomass, when combined with *Bacillus subtilis*. These findings underscore the potential of CuO-NPs and microbial inoculation for improving crop growth, which surpassed that of the control group in a field study [31]. These results highlight the potential of CuO nanoparticles and microbial inoculation for improving crop growth.

A study by Cao et al. [14] revealed that using copper oxide nanoparticles (CuO-NPs) can improve soil nitrogenase activity and increase diazotrophic diversity. However, CuO-NPs also led to a heightened centralization and complexity of the soil diazotrophic co-occurrence network, which destabilized the community structure. The assembly of the soil diazotrophic community was primarily driven by determinism, which remained consistent after treatment with CuO-NPs. This study suggested that CuO-NPs can shape diazotrophic communities, but their presence may lead to destabilization, potentially impacting crop nutrition and soil biological functions.

1.3.1 Copper-Based Nanocomposites

Polymer-Copper Nanocomposites

The incorporation of copper nanoparticles into polymers results in nanocomposites with enhanced mechanical, thermal, and antimicrobial properties. These materials have applications in packaging, biomedical devices, and coatings, demonstrating the synergistic effects of combining copper nanoparticles with polymeric matrices.

Carbon-Based Copper Nanomaterials

Hybrid materials comprising copper nanoparticles and carbon-based structures, such as carbon nanotubes or graphene, exhibit exceptional electrical conductivity and mechanical strength. These nanocomposites have been explored for electronics, energy storage, and catalysis applications.

Antimicrobial Properties: Copper-based nanomaterials' inherent antimicrobial properties make them valuable for applications in healthcare settings. They can inhibit the growth of bacteria and viruses, contributing to the development of antimicrobial surfaces and medical devices [81].

Catalysis and Sensing: Copper nanoparticles and nanocomposites are crucial in catalysis and sensing applications. Their high surface area, high reactivity, and tunable properties make them effective catalysts for various chemical reactions and sensitive components in sensors for detecting analytes.

The diverse types of copper-based nanomaterials and their unique properties make them promising candidates for a wide range of applications in various fields such as medicine, electronics, catalysis, and environmental remediation. As research in nanotechnology advances, further exploration of the synthesis techniques, properties, and applications of copper-based nanomaterials is essential for harnessing their full potential in various technological domains. Copper-based nanomaterials, including nanoparticles, nanocomposites, and nanowires, possess distinctive properties that make them promising candidates for postharvest applications. The antimicrobial activity, high surface area, and reactivity make them effective tools for controlling spoilage and extending the shelf life of harvested crops. Understanding these properties is crucial for harnessing their full potential in postharvest management.

1.4 Previous Studies on Copper-Based Nanomaterials in Agriculture

A comprehensive examination of the extant literature provides valuable insights into the utilization of copper-based nanomaterials within the agricultural domain. Previous investigations have systematically probed the antimicrobial efficacy, environmental ramifications, and crop-specific responses associated with these materials. The assimilation of knowledge accrued from antecedent research endeavors is pivotal for the meticulous design and effective implementation of strategies geared toward postharvest management.

The application of nanotechnology in agriculture, with a particular focus on copper-based nanomaterials, has attracted considerable attention due to its ability to augment crop yield, fortify soil health, and ameliorate disease control. This chapter endeavors to synthesize antecedent studies, providing a nuanced understanding of the diverse roles and impacts engendered by copper-based nanomaterials in the agricultural milieu. Copper (Cu) is a micronutrient crucial for numerous pivotal plant physiological processes. Its influence extends to essential functions such as photosynthesis, enzymatic action, and cellular processes, including cell wall development, oxidative stress response, mitochondrial respiration, and protein metabolism. As an indispensable micronutrient crucial for optimal crop development, copper nanoparticles (Cu-NPs) incorporating agrochemical agents, such as herbicides and fertilizers, are increasingly gaining traction in agriculture as viable substitutes for their traditional counterparts. As expected, the utilization of copper oxide nanoparticles (CuO-NPs) is projected to increase from 570 tons in 2014 to an estimated 1600 tons by the year 2025, as reported by Zhao et al. [90]. Furthermore, copper nanoparticles (Cu-NPs) are recognized for their pronounced antimicrobial characteristics, enhancing their agricultural efficacy in safeguarding crops from plant pathogens and diseases. In the North American region, an annual production volume of approximately 79,000 tons is dedicated to wood preservatives comprising aqueous copper carbonate, constituting a notable 50% share of the global market. These wood preservatives are identified explicitly for their incorporation of copper nanoparticles, which are characterized by a size range as small as 20 nm [9].

Scientific investigations have documented the efficacy of copper nanoparticles (Cu-NPs) as fungicides, demonstrating their potential utility against a diverse spectrum of plant fungi. These fungi include Fusarium sp., *Fusarium oxysporum, Curvularia lunata, Phoma destructiva, Alternaria alternate, Penicillium digitatum, Penicillium italicum*, and *Rhizoctonia solani.*

Copper nanoparticles have been utilized as nano-biofertilizers in cultivating *Cajanus cajan* L. (pigeon pea). Notably, Cu-NPs with dimensions ranging between 20 and 30 nm significantly enhanced in various growth parameters, including plant height, root length, weight, and germination rate of pigeon pea seedlings. [73].

Actinomycetal copper oxide nanoparticles play a crucial role in extending the shelf life of various food items, such as vegetables, including *Solanum lycopersicum.* The efficacy of microbial nanoparticles surpasses chemical preservatives, enhancing the longevity of vegetables such as *Solanum lycopersicum.* Microbial nanoparticles also have the potential to be beneficial for various other food products as well [15].

This application underscores the capacity of Cu-NPs to positively impact the overall growth and development of pigeon pea plants. Additionally, the controlled micronutrient release within agricultural systems was detected using nano-fertilizers, influencing both plant development and the biochemical processes governing crucial nutrients. Furthermore, varying concentrations of Cu-NPs were associated with increased productivity, better quality, elevated levels of enhanced levels of bioactive compounds, augmented protein content, and enhanced firmness in tomatoes. This observation highlights the multifaceted potential of Cu-NPs to influence diverse aspects of tomato plant development and yield positively. Furthermore, employing chitosan hydrogels encapsulating Cu-NPs revealed augmented tomato crop yields and enriched valuable compounds in the resulting tomato fruits.

Similarly, the effects of Cu-NPs on maize crops were investigated under both normal and stress conditions, demonstrating improved growth in normal conditions and reduced leaf curling and wilting during drought conditions. Additionally, maize plants subjected to stress conditions exhibited heightened pigmentation, increased yield, and improved water retention. These collective findings underscore the versatile potential of Cu-NPs, especially when encapsulated in chitosan hydrogels, in positively influencing the productivity and resilience of both tomato and maize crops. Similarly, an elevated wheat yield was documented upon the application of Cu-NPs. Under stress conditions, heightened reactive oxygen species (ROS) levels contribute to cellular injury affecting proteins and nucleic acids. Cu-NPs treatment has been associated with reduced ROS levels attributed to enhanced activity of superoxide dismutase (SOD) and ascorbate peroxidase (APx) enzymes, as evidenced in various studies. The introduction of copper to plants has demonstrated an increase in vitamin C content, a pivotal factor in photosynthesis and an enhancer of fruit quality.

Numerous reports underscore the augmentation of lycopene and flavonoid content and improvements in growth, yield, and shelf life across various plant species [11].

Hence, Cu-based nanoparticles offer numerous advantageous agricultural applications, from delivering agrochemicals to monitoring environmental pollutants. However, prolonged and continuous use of these nanoparticles in agriculture could pose various anticipated risks to the agro-ecosystem, especially the agricultural soil system. This encompasses their effects on soil micro and macrofauna, soil microbial populations, communities, diversity, and various soil processes like microbial interactions in soil cycles. Additionally, their effects on plants, rhizospheric interactions, food chain implications, and potential for trophic transfer are also concerning. Finally, their accumulation in soil and interactions with other soil components, including pollutants, could affect the overall quality and health of the soil system. In the realm of crop protection, copper nanoparticles have exhibited heightened efficacy as pesticides with concomitant reductions in environmental impact, attributable to their precision-targeted delivery systems that minimize the requisite quantity for efficacious pest control.

Furthermore, investigations reveal the potential utility of copper-based nanosensors for early disease detection in plants, affording real-time monitoring of subtle physiological changes and enabling timely interventions to preclude extensive crop damage. In nutrient management, copper-based nano-fertilizers, utilizing controlled release systems, have demonstrated enhanced nutrient use efficiency, mitigating environmental concerns related to conventional fertilizers by reducing nutrient runoff. Additionally, the integration of copper nanoparticles facilitates plants' superior nutrient uptake, with nano-sized carriers enhancing penetration through plant cell walls, thereby augmenting the absorption and utilization of essential elements. Nanoscale soil amendments, particularly copper-based nanomaterials, have proven effective in soil remediation endeavors by immobilizing or detoxifying contaminants, thereby enhancing soil health and mitigating the deleterious impacts of soil pollution. Nano-structured soil conditioners have enhanced soil structure, water retention, and nutrient availability, creating an environment conducive to optimal plant growth.

For precision agriculture, integrating nanoscale sensors facilitates real-time monitoring of soil conditions, water content, and crop health, fostering data-driven decision-making that optimizes resource utilization and enhances overall agricultural efficiency. Despite the promising prospects of copper-based nanomaterials in agriculture, concerns persist regarding their environmental impact, specifically in nanoparticle toxicity and persistence. Addressing these concerns is imperative for the responsible implementation of nanomaterials in agriculture. Regulatory and ethical challenges associated with using copper-based nanomaterials are also explored, highlighting the necessity for a balanced approach that harmonizes innovation with environmental and societal considerations to ensure sustainable implementation.

The amalgamation of antecedent studies underscores copper-based nanomaterials' multifaceted and promising role in agriculture. Nevertheless, the inherent challenges and ethical considerations underscore the imperative of sustained research efforts and judicious implementation to unlock their full potential within the purview of sustainable agricultural practices.

1.5 Significance of Copper-Based Nanomaterials

The significance of copper-based nanomaterials lies in their multifaceted applications in addressing postharvest challenges. From antimicrobial coatings to smart packaging, these materials offer a comprehensive approach to reducing losses and ensuring food safety. The potential economic, environmental, and health-related benefits underscore the importance of further exploring and harnessing these nanomaterials in agriculture.

1.5.1 Mechanism of Action

Both the foliar and root systems are often used to deliver nanoparticles to plants. These pathways utilize specific binding or carrier proteins, ion transport channels, or endocytosis to enter plant cells. Once particles enter the plant through the cuticles, trichomes, stomata, stigma, and hydathodes on the leaves, they are transported throughout the plant by means of phloem tissue. [64]. The transport of metal ions from the rhizosphere to the central vascular tissue is an active transport system that is involved in the absorption of metallic nanoparticles from the soil matrix. Through the plant's vascular system, CuO nanoparticles are taken up by the plant's upper sections from the root hairs. Although there has been little investigation into copper cellular transport, plants are known to contain heavy-metal transporters that collaborate with chelators and sequester transporters. Proteolytic enzyme, one of the plant cell membrane transporters involved in the movement of both beneficial and harmful metals is the copper transporter. The subgroups of big P-type ATPases are responsible for substrates pumping across biological membranes using ATP. Among the several copper transporters (COPT), COPT1 is a key player in the process of copper transport and cellular entrance. Nearly 140 thousand additionally, copper chaperones are crucial biological transporters that move copper from the cytosol to its final destinations. The plants have been shown to have three distinct copper chaperones: CCH, CCS, and COX17. Various proteins control the transcriptional level of all these metal nanoparticles (Tanaka 2018). An overabundance of copper was shown to stimulate mitogen-activated protein kinase (MAPS). It was discovered that MAPS has a role in phosphorylation and signal transduction [32].

1.5.2 Benefits and Advantages

The homeostatic management of copper (Cu), the eighth most prevalent metallic element in the Earth's crust, is tightly controlled since the element cannot be generated or destroyed once it has been dissolved [27]. The transition metal copper (Cu) has a metallic sheen, a noticeable reddish-orange hue, and an atomic number of 29, an atomic mass of 63.546, and a density higher than 5 g cm^{-3}. Impressive ductility, malleability, thermal and electrical conductivity, resistance to corrosion, and minimal chemical reactivity are just a few of its intriguing features. For countless generations, people have relied on copper's special qualities for a variety of vital social functions. Copper (Cu) is an essential trace element for all living things. A very little dosage (<100 mg/day) is required for human consumption (Fartusie et al. 2017). The fact that Cu may exist in either a reduced (Cu^{+}) or oxidized (Cu^{2+}) state via several routes is the trait that drives its many functions in structure and catalysis [89]. The interactions between metal ions and proteins have resulted in numerous biochemical reactions and structures, including the production of neuropeptides, regulating cell signaling pathways, bolstering antioxidant defense, activating immune cells such as macrophages, neutrophils, and helper T cells to combat pathogens, supporting immune system maintenance, and preserving the hair, elastic tissue in skin, bones, and other organs. This is due to the fact that Cu^{+} has an affinity for thioether groups, such as cysteine or methionine, and Cu^{2+} prefers coordination to oxygen or imidazole nitrogen groups, such as aspartic, glutamic acid, and histidine [20]. Optimal concentrations of Cu^{2+} and Cu^{+} in plants, in their native state, range from 10^{-14} to 10^{-16} M, respectively; contribute to the photosynthetic electron transport chain, mitochondrial respiration, response to oxidative stress, metabolism of the cell wall, hormone signaling, mobilization of iron, and other processes; act as a cofactor for multiple enzymes; and ensure the correct functioning of several important proteins and enzymes, including amino oxidase, cytochrome c oxidase, plastocyanin, and many more. At specific concentrations, it also forms the structural components of many regulatory proteins [66, 87].

1.5.3 Applications in Different Crops

Nanoparticles of copper are useful in farming since they are inexpensive and plants can absorb them easily due to their tiny size. Its antifungal, antibug, and antipest properties have been shown. Thus, it might be investigated for use as nanoscale fertilizers, insecticides, and herbicides, all of which have significant financial advantages. Instead of spending more money on each individual product, traditional farmers may save money by utilizing copper nanoparticles (Cu-NPs) as a fungicide, insecticide, micronutrient, and fertilizer all at once. As an antibacterial agent, copper nanoparticles are used in food packaging due to their affordability and environmental friendliness. It creates an impenetrable barrier that shields food from ultraviolet light, extending its storage life. A study found that copper nanoparticles increased drought tolerance and production efficiency by suppressing illnesses under stressful and dry circumstances. There have been reports that it may improve the levels of anthocyanin, chlorophyll, and carotenoids in plants, which in turn can boost seed output, resilience to stress, effluent control, and potential medicinal uses.

2 Agriculture Perspective: Cu-NPs as Antimicrobial Agents

The EPA reported that copper is an efficient antibacterial agent, lowering the microbial concentration responsible for severe infections by approximately 99.9%. Biosynthesized Cu-NPs ranging in size from 5 to 45 nm have potent antibacterial effects on several species of Staphylococcus and Enterococcus [5]. These nano-copper particles have also been shown to have antimicrobials effects on dangerous bacteria including *Bacillus subtilis, Vibrio cholerae, Staphylococcus aureus,* and *Pseudomonas aeruginosa* and have also been shown for these nano-copper particles. Variables such as temperature, pH, oxygen levels, and bacterial concentration determine the efficacy of this antibiotic (Kalińska et al. 2019). In addition to bacteria, Cu-NPs have been shown to inhibit viral, yeast, and fungal growth [78]. The gray mold fungal diseases caused by *Botrytis cinerea* significantly decrease the storage duration and marketing lifespan of tomatoes, posing a challenge for both growers and consumers. The postharvest loss of tomato was reduced by the use of green copper nanoparticles synthesized by using *Rosmarinus officinalis* extract [12]. The self-sanitizing properties of copper, as described by Grass [28], suggest that it might be used to destroy microbes at a high pace [28]. Copper nanoparticles emit a large amount of reactive oxygen species (ROS) that kill *E. coli* bacteria by oxidizing lipids, degrading DNA, and killing cells, Fig. 3. If the particles exhibit growth inhibition on liquid media, their antimicrobial activity may be detected by visualizing colony forming units (CFUs) or optical density (OD) on solid media [63], times scanning electron microscopy (SEM) reveals morphological alterations, including shape changes and the creation of cavities in cell walls, in microorganisms treated with Cu-NPs. Additionally, in regard to growth inhibition or prolonging the lag phase of microorganisms, the concentration of Cu-NPs is a major factor [51]. Furthermore, studies have suggested that electrical power could enhance the antimicrobial effectiveness of Cu-NPs in applications such as air filtration, water purification, and antibacterial packaging. One hundred fifty-two copper nanoparticles, made by reducing copper chloride with gelatin added as a stabilizer, showed enhanced antibacterial action against *E. coli* even at low concentrations [55]. Copper nanoparticles (Cu-NPs) produced by thermal decomposition exhibits approximately 85% free radical activity, which is significantly more than that of other metal oxide nanoparticles, and they have revealed powerful antibacterial and antioxidant properties [62]. Carbon nanotubes may enhance the surface area of nanoparticles, which has led to their application in the formation of multi-walled carbon nanotubes with Cu-NPs (MWCNT), according to research. By releasing copper ions that penetrate the cells of microbes and destroy them, this mixture significantly improves their antibacterial efficacy compared to pure Cu-NPs [76]. When compared to *Alternaria alternata* and *Fusarium solani, Penicillium chrysogenum* is the fungus species most impacted by Cu-NPs, although having the lowest minimum inhibitory concentration [59]. Overall, Cu-NPs increased the levels of organic carbon, total nitrogen, potassium, and phosphorus in tea plants while simultaneously increasing their antifungal action against red-root rot disease (Elango et al. 2015). Similarly, research has been conducted on the Colletotrichum fungus using nanoparticles of copper oxide

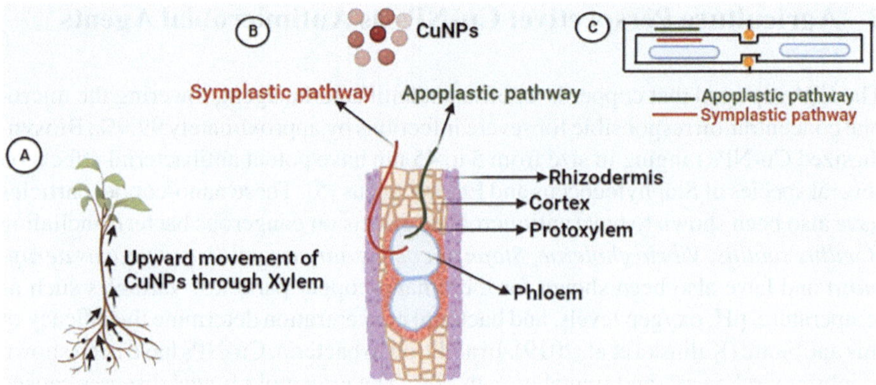

Fig. 2 Possible mechanisms for the uptake of copper nanoparticles in plants: **a** upward movement of Cu-NPs through the xylem, **b** uptake of Cu-NPs through root hairs into the vascular system, **c** two pathways for Cu-NPs movement within plant cells

and varying amounts of copper [57]. The antibacterial activity against *E. coli* and *Bacillus cereus* was demonstrated by the composite film consisting of Cu-NPs and chitosan, with iodine serving as a stabilizing agent [52]. The potential antibacterial properties of a nanocomposite film of copper nanoparticles (Cu-NPs) and cellulose have been demonstrated against *Staphylococcus aureus* and *Klebsiella pneumoniae* [61]. The findings of the antimicrobial investigation indicated that a composite film consisting of copper nanoparticles (Cu-NPs) and cellulose had significant efficacy against Pseudomonas sp., a bacterium known to cause food deterioration. Consequently, the film holds potential for application in food packaging [49] (Figs. 2 and 3).

2.1 Copper Nanoparticles Employed for Pest Control and Insect Management

Copper nanoparticles have a significant and beneficial impact on pest control in several plant species. The cotton leaf worm *Spodoptera littoralis* was able to lessen its impact on cotton (*Gossypium hirsutum*) plants when exposed to Cu-NPs at varying concentrations [1]. As the concentration of Cu-NPs increased, the death rate of *Tribolium castaneum* increased. The oxidative stress-inducing particles travel throughout the blood and lymphatic systems, passing via dendritic cells, epithelial cells, and endothelial cells [25, 33].

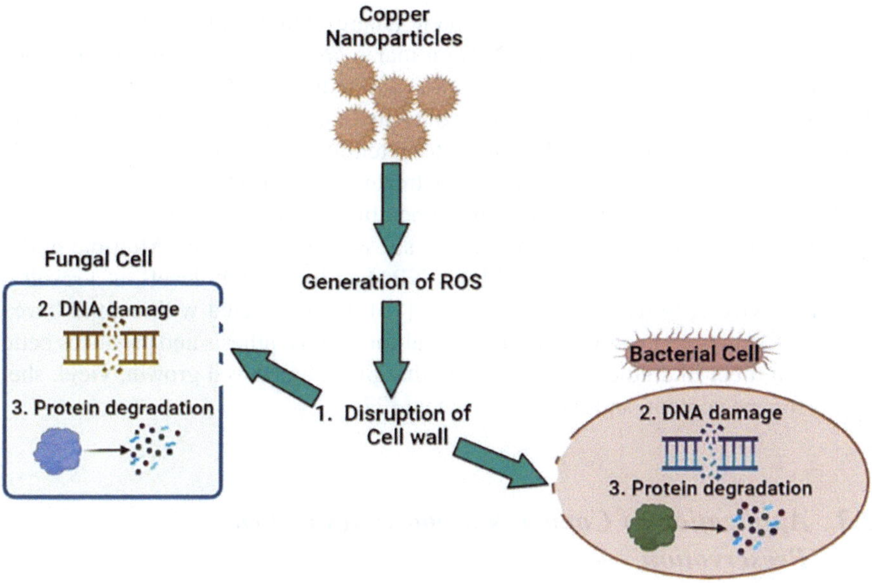

Fig. 3 Possible antimicrobial mechanisms of copper nanoparticles

2.2 Copper Nanoparticles Utilized as Plant Nutrient and Fertilizer

Plants in the agricultural systems rely on macro and micronutrients, which are necessary mineral elements, for the healthy development of their reproductive, vegetative, and other tissues. The macronutrients that plants need at greater quantities (>0.1% of dry tissue weight) include magnesium (Mg), potassium (K), nitrogen (N), calcium (Ca), sulfur (S), and phosphorus (P). In contrast, less than 0.01% of the dry tissue weight is needed for important micronutrients such as copper (Cu), nickel (Ni), iron (Fe), molybdenum (Mo), manganese (Mn), boron (B), zinc (Zn), and chlorine (Cl). Both macro and micronutrients play important roles in macromolecule structure and function. Copper is an essential micronutrient for plants; its abundance in chloroplasts attests to this. Copper is involved in many metabolic processes, including the production of chlorophyll and other pigments, as well as the metabolism of carbohydrates and proteins. *Cajanus cajan* L. (pigeon pea) growth was enhanced by applying Cu-NPs as nano-biofertilizers. The plant height, length, and weight of root, as well as pigeon pea seedlings germination were all enhanced by the 20–30 nm copper nanoparticles [73]. Furthermore, nano-fertilizers have been shown to impact plant development and nutrient biochemical cycles, allowing for the delayed release of micronutrients in agricultural systems. Improvement in yield, quality, bioactive chemicals, protein content, and tomatoes firmness were seen at various doses of Cu-NPs [2, 43]. Chitosan hydrogels encapsulating Cu-NPs increased tomato crop output and fruit beneficial chemical content (Juarez-Maldonado et al. 2016). Similarly, studies on the effects of Cu-NPs on maize crops have shown mixed results: increased growth under stress circumstances and reduced leaf curling and wilting in the absence of water. Under stressful circumstances, maize plants exhibit an increase in pigments, seed quantity, yield, and water retention, among other things over [37, 56]. Similarly, it was also shown that wheat treated with Cu-NPs increased yield [18]. Cellular proteins and nucleic acids are vulnerable to damage from reactive oxygen species (ROS), which are more abundant in stressful environments. Multiple studies have revealed that plants treated with Cu-NPs had lower ROS levels as a result of higher activity of SOD and APx enzymes [16]. Plants treated with copper have a higher vitamin C content, which is beneficial for photosynthesis and produces better fruit. A number of studies have indicated that plants improved growth, yield, shelf life, lycopene content, and flavonoid content [22].

2.3 Application of Copper Nanoparticles in Fruit Preservation

Fruit preservation is a critical aspect of the food industry to ensure extended shelf life and quality retention. Recent research has explored innovative approaches, with one notable avenue being the application of copper nanoparticles. This case study delves into the findings of a specific research project that investigated the use of copper nanoparticles for fruit preservation. This study addresses the prevention of microbial-induced food spoilage through the synthesis of copper sulfide nanoparticles (CuS-NPs) with inherent photothermal properties. The nanoparticles are successfully prepared via a one-step hydrothermal method and are subsequently uniformly dispersed within a natural carrageenan matrix to yield Carr/CuS films. Photothermal-induced antibacterial assessments reveal the broad-spectrum efficacy of the Carr/CuS film against *Escherichia coli* (99.2%) and *Staphylococcus aureus* (99.9%). Experimental validation on beef samples demonstrates a substantial enhancement in antibacterial performance upon near-infrared (NIR) irradiation. Furthermore, the Carr/CuS film, characterized by high transparency, exhibits superior mechanical strength and thermal stability compared to the Carr film. The incorporation of this NIR-induced strategy introduces a promising avenue for expanding the applicability of antibacterial packaging in food preservation [46].

Another research focused on the synthesis of antimicrobial copper nanoparticles incorporated into a cellulosic raw material comprising carboxymethyl cellulose and TEMPO-oxidized nano-fibrillated cellulose. The resulting hybrid material is subsequently integrated into polyvinyl alcohol (PVA) films. The antimicrobial efficacy of these hybrid nanocomposite films is evaluated against a non-pathogenic surrogate of the foodborne pathogen *Escherichia coli*. Additionally, the study provides insights into the size distribution, concentration, oxidation states, and release dynamics of copper nanoparticles from the nanocomposite films. The reinforcing impact of the hybrid nanocomposite material on PVA films, along with a concise examination of the thermal properties of these films, is also presented. Leveraging the well-established antimicrobial attributes of copper nanoparticles and the advantageous characteristics of cellulosic raw materials including renewability, high strength, low weight, substantial specific surface area, biocompatibility, and biodegradability, the resultant hybrid nanocomposite emerges as a promising bifunctional filler for applications in the food film packaging industry (Table 1).

Copper has been recognized for its potent toxicity against a broad spectrum of microorganisms. Different copper-based compounds have been developed for many bactericidal applications. The mechanism of copper toxicity to microorganisms remains to be fully understood and has been reported to be associated with copper interactions with the thiol groups of bacterial proteins and enzymes. Reports suggest that copper ions, released from the oxide layer of Cu-NPs, cause the disruption of certain lipopolysaccharide patches on the cell surface, resulting in changes to the permeability and functionality of the outer cell membrane. These ions have the ability to bind to DNA molecules, disrupting the helical structure through crosslinking within and between nucleic acid strands. Additionally, the production of reactive

Table 1 Nanoparticles composition showed the activities against different microorganisms

Nanoparticles	Particle size (nm)	Composition	Activity against	References
CuNPs	36 + 9 nm OR 36 − 9 nm	Polylactic acid (PLA)/ CuNPs	*P. fluorescens* *P. putida*	[50]
CuNPs	6 shapes and sizes	Agar/CuNPs	*L. monocytogenes* *E. coli*	[70]
ZnONPs	19.9 nm, 20.1 nm, 40.3 nm	Polypropylene (PP) packaging film	*S. aureus* *E. coli*	[58]
ZnONPs	75 nm	Poly (3ydroxybutyrate-co-3 hydroxy valerate) PHBV film	*S. aureus* *E. coli*	Diez-Pascual and Diez-Vicente (2014)
TiO_2NPs	7 nm, 5um	Oriented polypropylene (OPP) film	*E. coli*	Chawengkijwanich and Hayata (2008)
TiO_2NPs	7 nm	Polypropylene (PP)	*P. expansum*	[54]
TiO_2anatase	25 nm and 1 µm	HDPE + $CaCO_3$ TiO_2 composite matrix	Pseudomonas spp. Coliforms Lactic acid bacteria Yeast and molds	[29]

oxygen species (ROS) can result in irreversible damage, including protein oxidation and the fragmentation of DNA and RNA molecules [13].

Copper has long been recognized for its pronounced toxicity toward a diverse spectrum of microorganisms, prompting the development of various copper-based compounds for bactericidal applications [13]. Despite extensive research, a complete understanding of the mechanisms underlying copper toxicity to microorganisms still needs to be discovered. The current literature suggests a correlation between copper's impact and its interaction with thiol groups present in bacterial proteins and enzymes (Ladomersky and Petris 2015). The release of copper ions from the oxide layer of copper nanoparticles (Cu-NPs) is identified as a key contributor to microbial toxicity. This process is observed to trigger the disruption of patches of lipopolysaccharide on the bacterial cell surface, therefore modifying the permeability and functionality of the outer cell membrane. Copper ions further engage with molecules of DNA, inducing structural disarray through crosslinking within and among nucleic acid strands. Concurrently, the reactive oxygen species (ROS) generation during this interaction inflicts irreversible damage, including protein oxidation and DNA and RNA molecules cleavage [13].

The film composed of agar-based nanocomposites was fabricated through the combination of agar and Cu-NPs, which were manufactured utilizing three distinct salt types. The nanoparticles that were created were incorporated into packaging materials, specifically agar. The antibacterial properties of agar film were found to effectively prevent the growth of food spoilage bacteria [70]. Shankar et al. also documented that agar film with copper nanoparticles (Cu-NPs) can absorb UV light while maintaining its mechanical qualities and transparency. The FDA has advised using Ag–Cu bimetallic nanoparticles (0.5–4%) as a preventive measure against food deterioration [4]. The utilization of Cu-NP-embedded nanocomposite film in the packaging of food products has been demonstrated [76]. Similarly, the film composed of polyvinyl methyl ketone embedded with Cu-NP showed promising antibacterial properties and effectively inhibited the deterioration of food. In addition, the Cu-NP fluoropolymer coating effectively inhibits food deterioration [17].

2.4 Copper Oxide Nanomaterials for Grain Storage

An in-depth analysis was conducted on using copper oxide nanomaterials for grain storage. This study explored controlled release mechanisms, their impact on pest control, and comparisons with traditional storage methods. This case study critically evaluates the efficacy and feasibility of employing copper oxide nanomaterials in grain storage. Grain storage is critical to food security, and postharvest losses are a significant challenge. Recent research has explored using copper oxide nanomaterials as an innovative solution for grain storage. This case study delves into a specific research project that investigates the application of copper oxide nanomaterials in preserving grains. Commonly stored grains, such as wheat and rice, were treated with copper oxide nanomaterials. A control group of untreated grains was maintained for comparative analysis.

Extensive investigations have been conducted to elucidate the intricate association between *Oryza sativa* L. (rice) and copper oxide nanoparticles across diverse concentrations and experimental frameworks. The prevailing consensus in most of these studies substantiates a discernible impediment in rice growth upon CuO-NPs exposure significantly; the induction of oxidative stress by CuO nanoparticles has been recognized as a crucial factor contributing to the growth inhibition described earlier. Within the specific context of the Nippon bare cultivar, introduction of CuO-NPs elicited a discernible increase in lipid peroxidation within both root and shoot tissues. Importantly, the manifestation of this oxidative stress response has been ascertained to be under the regulatory influence of ethylene, thereby underscoring the intricate molecular mechanisms underpinning the observed physiological alterations in the plant [6]. In the Jyoti cultivar of rice, exposure for 30 days resulted in the detection of oxidative and osmotic stress, manifesting detrimental effects on productivity mainly attributed to structural impairment of the thylakoid membrane [19]. Concomitant studies, conducted by Da Costa and Sharma [19], yang et al. [85], and Rai et al. [65], have posited that compromised photosynthetic efficiency and apparatus integrity may constitute a significant contributory factor to the observed inhibition of growth. Notably, the influence of copper oxide nanoparticles (CuO-NPs) on rice appeared consistent across various experimental systems, with discernible suppression of seed germination and growth in various settings has been observed, including nanoparticle suspension on filter paper [80], soil [60], hydroponic systems [79], and cotton pads soaked in nano-CuO suspension [72]. Despite the unimpeded germination and shoot growth of rice seedlings, a significant reduction in the length of the root was still observed in the presence of CuO-NPs [84].

In contrast, investigations involving wheat (*Triticum aestivum* L.) as the investigational topic have yielded more complex and variable results. The impact of CuO-NPs on wheat growth is less uniform. Notably, the application of CuO-NPs has a favorable influence on wheat growth. Specifically, when administered through foliar spray or as grain pre-treatment, CuO nanoparticles showed a beneficial impact on the growth parameters of wheat plants, surpassing outcomes observed in the control group [8]. Moreover, the application of low concentrations of CuO-NP suspension exhibited increased growth and yield in wheat across hydroponic and in vitro systems [30, 86], as well as in soil [86]. Noteworthy findings by [39] elucidate concentration-dependent growth responses in wheat in vitro, where lower concentrations of CuO-NPs promoted plant growth, while increased concentrations lead to in a discernible reduction in plant growth. This nuanced perspective underscores the intricate and context-dependent nature of the interactions between CuO-NPs and wheat physiology. Microbiological assessments were conducted to quantify the reduced fungal and bacterial populations in treated grains. The study focused on prevalent storage fungi and bacteria that contribute to grain spoilage. The levels of mycotoxins, such as aflatoxins and ochratoxins, were analyzed in treated and untreated grains. High-performance liquid chromatography (HPLC) was employed to detect and quantify mycotoxin concentrations. The quality attributes of treated and untreated grains, including moisture content, germination rate, and nutritional content, were monitored over the storage period.

2.5 Copper Nanowires in Vegetable Packaging

Vegetable packaging is a critical aspect of the agricultural supply chain, influencing both the shelf life and produce quality. Recent advancements in nanotechnology have paved the way for innovative solutions to enhance packaging materials. This exploration focuses on the potential application of copper nanowires in vegetable packaging, aiming to extend shelf life, reduce spoilage, and contribute to sustainable agricultural practices. Due to their unique properties, copper nanowires offer a promising avenue for improving vegetable packaging. Copper exhibits inherent antimicrobial properties, preventing the growth of bacteria and fungi that contribute to spoilage. Due to the high electrical conductivity of copper nanowires, smart packaging with real-time monitoring capabilities is possible. Nanowires, with their nanoscale dimensions, provide flexibility and durability, making them suitable for incorporation into various packaging materials. Copper nanowires are synthesized through a solution-based method, utilizing copper salts and reducing agents. The resulting nanowires are characterized using electron microscopy and X-ray diffraction techniques to confirm their size, structure, and purity. Vegetable packaging materials are coated with a thin layer of copper nanowires. This coating can be applied to common packaging materials such as polyethylene or biodegradable polymers. The antimicrobial properties of copper nanowires act as a barrier against bacteria and fungi commonly associated with vegetable spoilage. This protective layer reduces the risk of microbial contamination during transportation and storage. Integration of copper nanowires enables the development of smart packaging with built-in sensors. These sensors monitor factors such as temperature, humidity, and gas concentrations inside the packaging, providing real-time data on the condition of the vegetables. The smart packaging system allows for continuous monitoring of vegetable quality. Changes in environmental conditions trigger alerts, enabling timely intervention to prevent spoilage and maintain freshness. Nanoparticles (NPs) synthesized from *Carica papaya* leaf extract using CuO as a catalyst exhibited the capability to degrade Coomassie brilliant blue dye R-250 under sunlight exposure [71, 88].

3 Challenges and Future Directions

As previously stated, tremendous progress has been made in this field; however, a number of difficulties still need to be overcome with the use of Cu-based nanomaterials. The difficulty of achieving reasonable structural control of Cu and Cu-based nanoparticles is a fundamental barrier to enhancing the material's functions and developing the desired shape, size, or defect-dependent characteristics. In addition, a thorough investigation is required to accurately characterize Cu-containing nanomaterials processed under various reaction conditions. Moreover, the potential environmental impact of copper nanoparticles in agriculture, especially regarding soil and water contamination, has raised concerns. It is essential to study the long-term effects and ecotoxicology to ensure sustainability. While copper is an essential micronutrient for plants, excessive amounts or the presence of copper nanoparticles may adversely affect non-target organisms in the ecosystem. Thus, it is important to understand the toxicity and impact of copper nanomaterials on beneficial organisms like soil microbes and earthworms. Also, ensuring plants' efficient uptake of copper nanoparticles without causing toxicity is challenging. The stability of copper nanoparticles is another concern, as the instability of copper nanoparticles in agricultural formulations limits their effectiveness. Thus, agglomeration and degradation over time reduce their efficacy, necessitating the development of stable formulations. Furthermore, the need for standardized regulations and guidelines for using copper nanoparticles in agriculture poses a challenge. Establishing clear regulatory frameworks is essential to ensure safe and responsible application. The biocompatibility of metal nanoparticles has gained attention for ensuring their non-toxic applications. Moreover, various biocompatible nanomaterials or copper nanoparticle coatings that can mitigate potential environmental and health risks have been explored. Thus, designing biocompatible and biodegradable nanoparticles that degrade into non-toxic components after fulfilling their function is a promising avenue for utilizing metal nanomaterials. Building awareness and fostering transparency will contribute to the responsible use of nanotechnology in farming practices. As the field of nanotechnology in agriculture evolves, addressing these challenges and pursuing these future directions will be essential to harness the full potential of copper nanoparticles while ensuring environmental sustainability and safety.

4 Conclusion

Various simple postharvest measures can be used to minimize postharvest losses, including appropriate harvest timing, washing and disinfection, classification, wrapping, precooling, and adequate transportation and distribution. Nevertheless, the efficacy of these techniques in preserving crop quality postharvest is only sometimes guaranteed. The synthesis of cost-effective nanomaterials with high efficiency is crucial for effectively managing postharvest illnesses and improving storage conditions while preserving crop quality. The efficacy of nanomaterials can be influenced by various factors, such as the species of crop, the dosage and type of nanomaterials used, the treatment timing, and the ripening stage. Copper nanoparticles induce degradation of the cell wall and disrupt the functionality of cellular organelles, ultimately resulting in the demise of infections. Hence, it can be utilized in the context of food packaging and food products to augment their longevity. To ensure the safer utilization of Cu-NPs in agriculture, it is imperative to establish a comprehensive understanding of the interactions between nanoparticles and plants. If implemented correctly, copper nanoparticles have the potential to significantly benefit agriculture and the food industry by enhancing the production and storage of food for extended durations while minimizing losses. To create standards and guarantee the safe use of these nanoparticles, it is essential to comprehend their toxicity and absorption processes. Nano-pesticides, nano-herbicides, and nano-fertilizers based on Cu-NPs can be developed as a new formulation for plant protection; these products are needed in lower quantities and could help reduce the toxicity caused by overuse of pesticides, as several studies have shown that Cu-NPs can fight fungi and insect-pests of crop plants. It is also possible to investigate Cu-NP-based biosensors for pest control and identifying food-spoiled diseases. Nevertheless, further research is needed into the exposure, toxicity, and ultimate destination of these Cu-NPs. Ultimately, to guarantee that Cu-NPs are safe for both the environment and living creatures, it is necessary to decrease the toxic effect of CuO-NPs to an insignificant level.

Competing Interests The authors declare that they have no competing interests.

Declaration of Generative AI and AI-Assisted Technologies in the Writing Process During the preparation of this work the book chapter author(s) have not taken any assistance from AI, and the book chapter author(s) reviewed and edited the content as needed and take(s) full responsibility for the content of the publication. All authors read and approved the final version of the manuscript.

References

1. Abdullah M, Numan M, Shafique MS, Shakoor A, Rehman S, Ahmad MI (2016) Genetic variability and interrelationship of various agronomic traits using correlation and path analysis in cotton (*Gossypium hirsutum* L.). Academia J Agric Res 4(6):315–318
2. Ahmed B, Khan MS, Musarrat J (2018) Toxicity assessment of metal oxide nano-pollutants on tomato (*Solanum lycopersicon*): a study on growth dynamics and plant cell death. Environ Pollut 240:802–816
3. Anyaogu KC, Fedorov AV, Neckers DC (2008) Synthesis, characterization, and antifouling potential of functionalized copper nanoparticles. Langmuir 24(8):4340–4346
4. Arfat YA, Ejaz M, Jacob H, Ahmed J (2017) Deciphering the potential of guar gum/Ag-Cu nanocomposite films as an active food packaging material. Carbohydr Polym 157:65–71
5. Arya A, Gupta K, Chundawat TS, Vaya D (2018) Biogenic synthesis of copper and silver nanoparticles using green alga *Botryococcus braunii* and its antimicrobial activity. In: Bioinorganic chemistry and applications
6. Azhar W, Khan AR, Salam A, Ulhassan Z, Qi J, Shah G, Gan Y (2023) Ethylene accelerates copper oxide nanoparticle-induced toxicity at physiological, biochemical, and ultrastructural levels in rice seedlings. Environ Sci Pollut Res 30(10):26137–26149
7. Bachheti RK, Abate L, Deepti, Bachheti A, Madhusudhan A, Husen H (2021) Algae-, fungi-, and yeast-mediated biological synthesis of nanoparticles and their various biomedical applications. In: Kharisov B, Kharissova O (eds) Handbook of greener synthesis of nanomaterials and compounds, vol 1: Fundamental principles and methods. Elsevier Inc. 50 Hampshire St., 5th Floor, Cambridge, MA 02139, USA, pp 701–734. https://doi.org/10.1016/B978-0-12-821938-6.00022-0

8. Badawy AA, Abdelfattah NA, Salem SS, Awad MF, Fouda A (2021) Efficacy assessment of biosynthesized copper oxide nanoparticles (CuO-NPs) on stored grain insects and their impacts on morphological and physiological traits of wheat (*Triticum aestivum* L.). Plant Biol 10:233
9. Bakshi M, Kumar A (2021) Copper-based nanoparticles in the soil-plant environment: assessing their applications, interactions, fate, and toxicity. Chemosphere 281:130940
10. Bantang JPO, Bigol UG, Camacho DH (2021) Gel and film composites of silver nanoparticles in κ-, ι-, and λ-Carrageenans: one-pot synthesis, characterization, and bioactivities. Bionanoscience 11:53–66
11. Bhagat M, Anand R, Sharma P, Rajput P, Sharma N, Singh K (2021) Review-multifunctional copper nanoparticles: synthesis and applications. ECS J Solid-State Sci Technol 10(6):063011
12. Bikdeloo M, Ahsani Irvani M, Roosta HR, Ghanbari D (2021) Green synthesis of copper nanoparticles using rosemary extract to reduce postharvest decays caused by Botrytis Cinerea in tomato. J Nanostruct 11(4):834–841
13. Borkow G, Gabbay J (2009) Copper, an ancient remedy returning to fight microbial, fungal and viral infections. Curr Chem Biol 3(3):272–278
14. Cao J, Liu Z, Zhao H, Lai F, Han Y, Lin X (2023) Effects of copper oxide nanoparticles on soil diazotrophic communities in maize rhizosphere. J Soils Sediments 23(4):1760–1774
15. Chaugule VV (2021) Microbial synthesis of copper oxide nanoparticles from endophytic actinomycetes and its postharvest application on *Solanum lycopersicum*. J Postharvest Technol 9(3):68–83
16. Choudhury FK, Rivero RM, Blumwald E, Mittler R (2017) Reactive oxygen species, abiotic stress and stress combination. Plant J 90(5):856–867
17. Cioffi N, Ditaranto N (2005) Analytical characterization of bioactive fluropolymer ultra-thin coatings modified by copper nanoparticles. Anal Bioanal Chem 381:607–616
18. Cota-Ruiz K, Hernández-Viezcas JA, Varela-Ramírez A, Valdés C, Núñez-Gastélum JA, Martínez-Martínez A, Gardea-Torresdey JL (2018) Toxicity of copper hydroxide nanoparticles, bulk copper hydroxide, and ionic copper to alfalfa plants: a spectroscopic and gene expression study. Environ Pollut 243:703–712
19. Da Costa MVJ, Kevat N, Sharma PK (2020) Copper oxide nanoparticle and copper (II) ion exposure in *Oryza sativa* reveals two different mechanisms of toxicity. Water Air Soil Pollut 231:1–16
20. Dey A, Manna S, Kumar S, Chattopadhyay S, Saha B, Roy S (2020) Immunostimulatory effect of chitosan conjugated green copper oxide nanoparticles in tumor immunotherapy. Cytokine 127:154958
21. Doria G, Conde J, Veigas B, Giestas L, Almeida C, Assunção M, Rosa J, Baptista PV (2012) Noble metal nanoparticles for biosensing applications. Sensors 12(2):1657–1687
22. Dutta, T, Neelapu NRR, Surekha C (2020) Iron, zinc, and copper application in overcoming environmental stress. In: Protective chemical agents in the amelioration of plant abiotic stress: biochemical and molecular perspectives, pp 582–596
23. Eastman J, Choi S, Li S, Yu W, Thompson L (2001) Anomalously increased effective thermal conductivity of ethylene glycol-based nanofluids containing copper nanoparticles. Appl Phys Lett 78:718–720
24. El-Saadony MT, El-Hack A, Mohamed E, Taha AE, Fouda MM, Ajarem JS (2020) Ecofriendly synthesis and insecticidal application of copper nanoparticles against the storage pest *Tribolium castaneum*. Nanomaterials 10(3):587
25. Engin AB, Nikitovic D, Neagu M, Henrich-Noack P, Docea AO, Shtilman MI, Tsatsakis AM (2017) Mechanistic understanding of nanoparticles' interactions with extracellular matrix: the cell and immune system. Part Fibre Toxicol 14:1–16
26. Feregrino-Perez AA, Magaña-López E, Guzmán C, Esquivel K (2018) A general overview of the benefits and possible adverse effects of the nanotechnology in horticulture. Sci Hortic 238:126–137
27. Gawande MB, Goswami A, Felpin FX, Asefa T, Huang X, Silva R, Varma RS (2016) Cu and Cu-based nanoparticles: synthesis and applications in catalysis. Chem Rev 116(6):3722–3811

28. Grass G, Rensing C, Solioz M (2011) Metallic copper as an antimicrobial surface. Appl Environ Microbiol 77(5):1541–1547
29. Gumiero M, Peressini D, Pizzariello A, Sensidoni A, Iacumin L, Comi G, Toniolo R (2013) Effect of TiO_2 photocatalytic activity in a HDPE-based food packaging on the structural and microbiological stability of a short-ripened cheese. Food Chem 1:138
30. Hafeez A, Razzaq A, Mahmood T, Jhanzab HM (2015) Potential of copper nanoparticles to increase growth and yield of wheat. J Nanosci Adv Technol 1(1):6–11
31. Haider HI, Zafar I, Ain QU, Noreen A, Nazir A, Javed R, Sehgal SA, Khan AA, Rahman MM, Rashid S, Garai S, Sharma R (2022) Synthesis and characterization of copper oxide nanoparticles: its influence on corn (Z. mays) and wheat (*Triticum aestivum*) plants by inoculation of *Bacillus subtilis*. Environ Sci Pollution Res 30(13):37370–37385
32. Hendriks IA, Lyon D, Young C, Jensen LJ, Vertegaal AC, Nielsen ML (2017) Site-specific mapping of the human SUMO proteome reveals co-modification with phosphorylation. Nat Struct Mol Biol 24(3):325–336
33. Hong J, Zhang YQ (2016) Murine liver damage caused by exposure to nano-titanium dioxide. Nanotechnology 27(11):112001
34. Hou L, Li K, Ding Y, Li Y, Chen J, Wu X, Li X (2012) Removal of silver nanoparticles in simulated wastewater treatment processes and its impact on COD and NH_4 reduction. Chemosphere 87(3):248–252
35. Husen A (2022) Engineered nanomaterials for sustainable agricultural production, soil improvement and stress management. Elsevier Inc., 50 Hampshire St., 5th Floor, Cambridge, MA 02139, USA. https://doi.org/10.1016/C2021-0-00054-7
36. Husen A (2025) Emerging carbon nanomaterials for sustainable agricultural practices—synthesis, plant growth, performance, production and protection. Springer Nature Singapore Pte Ltd., 152 Beach Road, #21-01/04 Gateway East, Singapore 189721, Singapore
37. Hussain A, Rizwan M, Ali Q, Ali S (2019) Seed priming with silicon nanoparticles improved the biomass and yield while reduced the oxidative stress and cadmium concentration in wheat grains. Environ Sci Pollut Res 26:7579–7588
38. Hussain I, Singh NB, Singh A, Singh H, Singh SC (2016) Green synthesis of nanoparticles and its potential application. Biotech Lett 38:545–560
39. Ibrahim AS, Ali GA, Hassanein A, Attia AM, Marzouk ER (2022) Toxicity and uptake of CuO nanoparticles: evaluation of an emerging nano fertilizer on wheat (*Triticum aestivum* L.) plant. Sustainability 14(9):4914
40. Joshi A, Sharma A, Bachheti RK, Husen A, Mishra VK (2019) Plant-mediated synthesis of copper oxide nanoparticles and their biological applications. In: Nanomaterials and plant potential. Springer, Cham, pp 221–237. https://doi.org/10.1007/978-3-030-05569-1_8
41. Kamle M, Mahato DK, Devi S, Soni R, Tripathi V, Mishra AK, Kumar P (2020) Nanotechnological interventions for plant health improvement and sustainable agriculture. Biotech 10:1–11
42. Khot LR, Sankaran S, Maja JM, Ehsani R, Schuster EW (2012) Applications of nanomaterials in agricultural production and crop protection: a review. Crop Prot 35:64–70
43. Klunklin W, Savage G (2017) Effect on quality characteristics of tomatoes grown under well-watered and drought stress conditions. Foods 6(8):56
44. Kruis FE, Fissan H, Peled A (1998) Synthesis of nanoparticles in the gas phase for electronic, optical and magnetic applications—a review. J Aerosol Sci 29:511–535
45. Kumar A, Choudhary A, Kaur H, Mehta S, Husen A (2021) Metal-based nanoparticles, sensors and their multifaceted application in food packaging. J Nanobiotechnol 19(256):1–25. ISBN: 1477-3155. https://doi.org/10.1186/s12951-021-00996-0
46. Li F, Liu Y, Cao Y, Zhang Y, Zhe T, Guo Z, Sun X, Wang Q, Wang L (2020) Copper sulphide nanoparticle-carrageenan films for packaging application. Food Hydrocolloids 109:106094
47. Liu Y, Li Y, Schanze KS (2002) Photophysics of π-conjugated oligomers and polymers that contain transition metal complexes. J Photochem Photobiol C 3:1–23
48. Lobregas MOS, Bantang JPO, Camacho DH (2019) Carrageenan-stabilized silver nanoparticle gel probe kit for colorimetric sensing of mercury (II) using digital image analysis. Sensing Bio-Sensing Res 26:100303

49. Longano D, Ditaranto N, Di Niso F, Sibillano T, Ancona A, Conte A, Del Nobile MA, Sabbatini L, Torsi L (2012) Analytical characterization of laser-generated copper nanoparticles for antibacterial composite food packaging. Anal Bioanal Chem 403:1179–1186
50. Longano DN, Ditaranto N, Cioffi F, Di Niso T, Sibillano A, Ancona A, Conte MA, Del Nobile L, Torsi SL (2012) Analytical characterization of laser-generated copper nanoparticles for antibacterial composite food packaging. Anal Bioanal Chem 403:1179–1186
51. Ma Y, Huang R, Qi W, Su R, He Z (2019) Fluorescent silicon nanoparticles inhibit the amyloid fibrillation of insulin. J Mater Chem B 7(9):1397–1403
52. Mallick S, Sharma S, Banerjee M, Ghosh S, Chattopadhyay A, Paul A (2012) Iodine-stabilized Cu nanoparticle chitosan composite for antibacterial applications. ACS Appl Mater Interfaces 4:1313–1323
53. Mallik K, Mandal M, Pradhan N, Pal T (2001) Seed mediated formation of bimetallic nanoparticles by UV irradiation: a photochemical approach for the preparation of "Core–Shell" type structures. Nano Lett 1:319–322
54. Maneerat C, Hayata Y (2006) Antifungal activity of TiO_2 photocatalysis against *Penicillium expansum* in vitro and in fruit tests. Int J Food Microbiol 107:99–103
55. Morsi RE, Alsabagh AM, Nasr SA, Zaki MM (2017) Multifunctional nanocomposites of chitosan, silver nanoparticles, copper nanoparticles and carbon nanotubes for water treatment: antimicrobial characteristics. Int J Biol Macromol 97:264–269
56. Nakabayashi R, Yonekura-Sakakibara K, Urano K, Suzuki M, Yamada Y, Nishizawa T, Saito K (2014) Enhancement of oxidative and drought tolerance in Arabidopsis by overaccumulation of antioxidant flavonoids. Plant J 77(3):367–379
57. Oussou-Azo AF, Nakama T, Nakamura M, Futagami T, Vestergaard MDCM (2020) Antifungal potential of nanostructured crystalline copper and its oxide forms. Nanomaterials 10(5):1003
58. Paisoonsin S, Pornsunthorntawee O, Rujiravanit R (2013) Preparation and characterization of ZnO-deposited DBD plasma-treated PP packaging film with antibacterial activities. Appl Surf Sci 273:824–835
59. Pariona N, Mtz-Enriquez AI, Sanchez-Rangel D, Carrion G, Paraguay-Delgado F, Rosas-Saito G (2019) Green-synthesized copper nanoparticles as a potential antifungal against plant pathogens. RSC Adv 9:18835–18843
60. Peng C, Xu C, Liu Q, Sun L, Luo Y, Shi J (2017) Fate and transformation of CuO nanoparticles in the soil–rice system during the life cycle of rice plants. Environ Sci Technol 51(9):4907–4917. https://doi.org/10.1021/acs.est.6b05882
61. Pinto RJB, Daina S, Sadocco P, Neto CP, Trindade T (2013) Antibacterial activity of nanocomposites of copper and cellulose. BioMed Res Int 280512
62. Rabiee N, Bagherzadeh M, Ghadiri AM, Kiani M, Aldhaher A, Ramakrishna S, Webster TJ (2020) Green synthesis of ZnO NPs via *Salvia hispanica*: evaluation of potential antioxidant, antibacterial, mammalian cell viability, H1N1 influenza virus inhibition and photocatalytic activities. J Biomed Nanotechnol 16(4):456–466
63. Raffi M, Mehrwan S, Bhatti TM, Akhter JI, Hameed A, Yawar W, ul Hasan MM (2010) Investigations into the antibacterial behavior of copper nanoparticles against *Escherichia coli*. Ann Microbiol 60:75–80
64. Rai M, Ingle AP, Pandit R, Paralikar P, Shende S, Gupta I, da Silva SS (2018) Copper and copper nanoparticles: role in management of insect-pests and pathogenic microbes. Nanotechnol Rev 7(4):303–315
65. Rai P, Singh VP, Peralta-Videa J, Tripathi DK, Sharma S, Corpas FJ (2021) Hydrogen sulfide (H2S) underpins the beneficial silicon effects against the copper oxide nanoparticles (CuO NPs) phytotoxicity in *Oryza sativa* seedlings. J Hazard Mater 415:124907
66. Rajput V, Minkina T, Ahmed B, Sushkova S, Singh R, Soldatov M, Gorovtsov A (2020) Interaction of copper-based nanoparticles to soil, terrestrial, and aquatic systems: critical review of the state of the science and future perspectives. Rev Environ Contam Toxicol 252:51–96
67. Rajput VD, Minkina T, Ahmed B, Singh VK, Mandzhieva S, Sushkova S, Wang B (2022) Nanobiochar: a novel solution for sustainable agriculture and environmental remediation. Environ Res 210:112891

68. Rosbero TM, Camacho DH (2017) Green preparation and characterization of tentacle-like silver/copper nanoparticles for catalytic degradation of toxic chlorpyrifos in water. J Environ Chem Eng 5:2524–2532
69. Sanguansri P, Augustin MA (2006) Nanoscale materials development—a food industry perspective. Trends Food Sci Technol 17(10):547–556
70. Shankar S, Teng X, Rhim JW (2014) Properties and characterization of agar/CuNP bionanocomposite films prepared with different copper salts and reducing agents. Carbohyd Polym 114:484–492
71. Sharma A, Bachheti A, Sharma P, Bachheti RK, Husen A (2020) Phytochemistry, pharmacological activities, nanoparticle fabrication, commercial products and waste utilization of *Carica papaya* L.: a comprehensive review. Curr Res Biotechnol 2:145–160
72. Shaw AK, Hossain Z (2013) Impact of nano-CuO stress on rice (*Oryza sativa* L.) seedlings. Chemosphere 93(6):906–915
73. Shende S, Rathod D, Gade A, Rai M (2017) Biogenic copper nanoparticles promote the growth of pigeon pea (*Cajanus cajan* L.). IET Nanobiotechnol 11(7):773–781
74. Singh N, Gaur S, Chawla S, Singh S, Husen A (2023) Use of smart nanomaterials in food packaging. In: Husen A, Siddiqi KS (eds) Advances in smart nanomaterials and their applications. Elsevier Inc., 50 Hampshire St., 5th Floor, Cambridge, MA 02139, USA, pp 223–245 ISBN: 978-0-323-99546-7
75. Sytu MRC, Camacho DH (2018) Green synthesis of silver nanoparticles (AgNPs) from *Lenzites betulina* and the potential synergistic effect of AgNP and capping biomolecules in enhancing antioxidant activity. BioNanoScience 8:835–844
76. Tamayo L, Azócar M, Kogan M, Riveros A, Páez M (2016) Copper-polymer nanocomposites: an excellent and cost-effective biocide for use on antibacterial surfaces. Mater Sci Eng C 69:1391–1409
77. Thakkar KN, Mhatre SS, Parikh RY (2010) Biological synthesis of metallic nanoparticles. Nanomedicine 6(2):257–262
78. Vincent M, Duval RE, Hartemann P, Engels-Deutsch M (2018) Contact killing and antimicrobial properties of copper. J Appl Microbiol 124(5):1032–1046
79. Wang S, Liu H, Zhang Y, Xin H (2015) The effect of CuO NPs on reactive oxygen species and cell cycle gene expression in roots of rice. Environ Toxicol Chem 34(3):554–561
80. Wang W, Liu J, Ren Y, Zhang L, Xue Y, Zhang L, He J (2020) Phytotoxicity assessment of copper oxide nanoparticles on the germination, early seedling growth, and physiological responses in *Oryza sativa* L. Bull Environ Contam Toxicol 104:770–777
81. Wassie AT, Ahmed IN, Bachheti A, Husen A, Bachheti RK (2024) Plant extract-assisted green synthesis and characterization of copper nanocomposite. In: Metal and metal-oxide based nanomaterials: synthesis, agricultural, biomedical and environmental interventions. Springer Nature Singapore, Singapore, pp 43–66. https://doi.org/10.1007/978-981-99-7673-7_3
82. Worku LA, Bachheti A, Chaubey KK, Bachheti RK, Husen A (2023a) Antimicrobial activities of nanoemulsion. In: Husen A, Bachheti RK, Bachheti A (eds) Current trends in green nanoemulsions. Springer Nature Singapore Pte Ltd., 152 Beach Road, #21-01/04 Gateway East, Singapore 189721, Singapore, pp 143–156. https://doi.org/10.1007/978-981-99-5398-1_8
83. Worku LA, Deepti, Nigussie Y, Bachheti A, Bachheti RK, Husen A (2023b) Antimicrobial activities of nanomaterials. In: Husen A, Siddiqi KS (eds) Advances in smart nanomaterials and their applications. Elsevier Inc., 50 Hampshire St., 5th Floor, Cambridge, MA 02139, USA, pp 127–148
84. Yang Z, Chen J, Dou R, Gao X, Mao C, Wang L (2015) Assessment of the phytotoxicity of metal oxide nanoparticles on two crop plants, maize (*Zea mays* L.) and rice (*Oryza sativa* L.). Int J Environ Res Public Health 12(12):15100–15109
85. Yang Z, Xiao Y, Jiao T, Zhang Y, Chen J, Gao Y (2020) Effects of copper oxide nanoparticles on the growth of rice (*Oryza sativa* L.) seedlings and the relevant physiological responses. Int J Environ Res Public Health 17(4):1260
86. Yasmeen F, Raja NI, Ilyas N, Komatsu S (2018) Quantitative proteomic analysis of shoot in stress tolerant wheat varieties on copper nanoparticle exposure. Pl Mol Bio Rep 36:326–340

87. Yruela I (2009) Copper in plants: acquisition, transport and interactions. Funct Plant Biol 36(5):409–430
88. Zamindar N, Anari ES, Bathaei SS, Shirani N, Tabatabaei L, Mahdavi-Asl N, Khalili A, Paidari S (2020) Application of copper nano particles in antimicrobial packaging: a mini review. Act Sci Nutr Health 4:14–18
89. Zhang L, Webster TJ (2009) Nanotechnology and nanomaterials: promises for improved tissue regeneration. Nano Today 4(1):66–80
90. Zhao S, Su X, Wang Y, Yang X, Bi M, He Q, Chen Y (2020) Copper oxide nanoparticles inhibited denitrifying enzymes and electron transport system activities to influence soil denitrification and N_2O emission. Chemosphere 245:125394

Use of Copper Nanomaterials Against Nematodes

Subhrajit Banerjee, Sayan Kar, and Kasturi Sarkar

Abstract Nanoparticles (NPs) have recently been used in agriculture to cure soil of its polluted state and function as a pesticide because of their non-toxic and eco-friendly nature. Phytonematodes are significant causes of crop loss all over the world. The chapter encircles different reports of copper nanoparticles, which are biodegradable, water-soluble, permeable, and thermally stable and have the potential to function as an effective nematocidal agent in plant pathologies. NPs act differently than elemental copper in their nano form. Nanoformulation of copper improves its activity against nematodes. Different forms of nanoparticles, including encapsulated, caged, and nano forms, are reported to be used in agriculture as nematicides. The potential of copper nanoparticles, both inorganic and organically derived metallic oxide NPs having nematicidal activity has been discussed.

Keywords Copper nanoparticles · Nematicidal effects · Green biosynthesized nanoparticles · Environmental effects

1 Introduction

1.1 Nanoparticles and Their Applications

Nanoparticles (NPs) are ultrafine polymeric particles with less than 100 nm dimensions. Several types of NPs are based on size, shape, and physicochemical properties [1]. Depending on their dimensions, NPs are classified as 0-dimensional (D), 1D, 2D, or 3D [2]. The optical properties of the NPs are dependent on their size. Based

S. Banerjee · S. Kar · K. Sarkar (✉)
Department of Microbiology, St. Xavier's College, Kolkata, India
e-mail: sarkar.kasturi@sxccal.edu

S. Banerjee
e-mail: subhrajitbanerjee@surendranathcollege.ac.in

S. Banerjee
Department of Physiology, Surendranath College, Kolkata, India

R. K. Bachheti et al. (eds.), *Copper-Based Nanomaterial for Agricultural Practices*, Smart Nanomaterials Technology, https://doi.org/10.1007/978-981-96-8610-0_11

on the material they produce, NPs can be carbon-based, metal, ceramic, semiconductor, and polymeric. Biodegradable NPs [3] have gained interest as effective drug delivery systems in the last few decades. The semiconductor and metallic NPs have huge applications in diagnosing diseases like cancer [4] and hyperglycemia [5] due to their surface plasmon resonance-enhanced light scattering and absorption. As lubricants, adhesives, and coating materials, NPs can be applied in mechanical industries. The broad absorption band of alkali and noble metals in the visible zone of the electromagnetic solar spectrum has found wide applications [6]. Copper-based NPs (CuNPs) are used in applications like catalytic organic transformations, electrocatalysis, and photocatalysis. However, the use of various NPs and nanomaterials and their destruction response should also be monitored as specific toxicities are associated with them [7]. It is proposed that NPs of < 10 nm act like a gas and can enter any organism through inhalation or ingestion of food and translocate to organs and tissues where they can exert toxicological effects [8, 9]. In agricultural fields, NPs can cure the soil by removing nematodes, but a high concentration of NPs can harm the plants. The application of copper oxide nanoparticles (CuONPs) at a concentration of 450 mg/L has been found to cause increased ROS production, leading to cellular damage and less plant biomass accumulation in rice seedlings [10]. The addition of ethylene, the gaseous phytohormone, to plants increases the toxic effects of CuONPs even more [10].

In this chapter, we will discuss the nematicidal effects of CuNPs, both green and chemically synthesized, on the common nematode *Meloidogyne incognita.* We will also assess the toxicological effects of NPs on *Caenorhabditis elegans,* as found in recent papers.

2 Historical Evidence of the Use of Nanoparticles

CuNPs had been used to color and impart glaze to clay potteries in Mesopotamia since the ninth century. Cu and silver (Ag) salts were added to the clay used to make potteries. At the time of baking under conditions like elevated temperatures and a reducing environment, some of the metal ions moved to the outer layer of the pottery and were reduced to metals, creating two layers of CuNPs. On exposure to light, some light is reflected by the outer layer, and some penetrates inside and is reflected by the second layer. The light emitted from both layers causes constructive and destructive interference, creating a luster effect [11].

3 Mechanism of Toxicity of Cu and CuNP

Copper surfaces are known to reduce the microbial load and have been used for a long time for this purpose. Cu, being a 3d transition metal, can attain a wide range of oxidation states (0, 1, 2, 3) and can react via both one- and two-electron pathways.

The cellular toxicity effects of Cu are not fully known, but it affects multiple cellular targets and causes rupture of cellular membranes, accumulation of ions inside the cell, unfolding and inactivation of proteins, inhibition of enzymes, displacement of Fe–S centers, damage to DNA, and ROS production.

CuNPs are either metallic Cu or copper oxide (CuO) NPs. Their properties vary depending on their physicochemical properties, like dissolution, aggregation behavior, and ROS production. CuNPs have unique characteristics that are not observed in Cu metals. They show extraordinarily catalytic activity due to a large surface-to-volume ratio. Smaller sizes with great porosity help achieve a higher reaction yield quickly [12]. Both CuNPs and CuONPs act against worms, crustacean species, bacteria, algae, and human cell lines [13]. As these NPs oxidize slowly in solutions, Cu^{+2} ions are released, which create toxic hydroxyl free radicals (*OH). These free radicals interact with the microbial lipid membrane and degenerate, causing the death of the microorganism. CuONPs also enter the mitochondria and trigger ROS production. CuONPs enter the nucleus through nuclear pores or damaged nuclear membranes, releasing more Cu^{2+}, leading to oxidative damage and DNA damage [14]. CuNPs interact with proteins, unfolding proteins, denaturation, thiol crosslinking, and loss of function [15, 16].

4 Use of CuNPs in Agriculture

Nanotechnology has progressed slowly in agriculture compared to other industries because of the difficulty and lack of field trials. There are two ways of protecting or managing plant losses using NPs. They can be used alone to kill pests, nematodes, and bacteria or as carriers for insecticides, fungicides, herbicides, and RNA-interference molecules.

Cu is essential in micro amounts in photosynthesis, enzymatic activity, and various other cellular functions. Also, it is known to function as an antimicrobial and fungicide [17–19]. The United States Environmental Protection Agency (EPA) recognized Cu and its alloys as the first metallic antimicrobial agent in 2008 [20]. CuNPs are even more toxic and show differential effects than metal Cu ions [21, 22]. Moreover, Cu is abundant in nature and less expensive. Hence, CuNPs containing herbicides/fertilizers are gaining popularity for their beneficial effects over their traditional chemical counterparts [23]. An estimation says that the production of CuONPs would be 1600 tons in 2025 [24].

5 Phytonematodes

Phytonematodes, or plant-parasitic nematodes, are plant pathogens smaller than 1 mm that feed on plant tissues [25]. Phytonematodes infect a wide range of host plants, including major agricultural plants and crops, leading to severe losses yearly

[26]. Among the phytonematodes, the most prominent and damaging ones are the root-knot nematodes (RKN), which belong to *the Meloidogyne* species. The other nematode species include the ring nematodes, root lesion nematodes, sugar beet cyst nematodes, citrus nematodes, and stem and bulb nematodes. The major phytonematode *Meloidogyne incognita,* belonging to the family Heteroderidae, is a very common species worldwide and can infect many host plants. It is not easy to control *Meloidogyne incognita,* as it spreads quickly from garden to garden through infested plants, soil, and tools. *M. incognita* follows thermotaxis to move toward an appropriate level in soil and searches for chemical cues to reach specific plant roots [27]. *Meloidogyne incognita* attacks more than 3000 plant species, including both monocotyledons and dicotyledons, and is responsible for severe crop loss.

The entire lifecycle of *Meloidogyne incognita* takes 37 days to complete at 21 °C. The first molt occurs inside the egg, and the juvenile first stage (J1) is converted to the juvenile second stage (J2). With the help of favorable conditions like temperature, moisture, and other stimuli, the J2 hatches and infects the host plant's root. Upon invading the roots, J2 moves through the cortical tissues toward the vascular zone and creates a permanent feeding site where it enlarges and forms a giant cell. Eventually, J2 matures, becomes J3 and J4, and finally transforms into an adult nematode that leaves the root. Different body shapes, like vermiform shapes for males and globose-shaped bodies for females, help to identify male and female *M. incognita.* The eggs laid by females initiate the life cycle of *M. incognita* [25, 28, 29].

Another common and popular model nematode for experimentation is *Caenorhabditis elegans.* It was first used as a model organism to study animal development by Sydney Brenner in 1965 [30, 31]. Its short lifespan (3 days), small size (1.5 mm), easy cultivation in the laboratory, and transparent body make it possible to observe the changes inside the nematode using simple microscopy or more advanced techniques. *C. elegans* is a good model for neural development due to the presence of only 302 neurons and 7000 synapses. The genetic map containing more than 100 genetic loci, which carry mainly behavioral or morphological markers, is distributed over six chromosomes. *C.elegans*, being a eutelic organism, shows constancy in cell number and cell position from individual to individual, which helps in developmental studies. The complete development of adult *C.elegans* from fertilized egg through cell divisions and migrations can be observed in living animals [32]. As the use of nanomaterials and NPs is increasing and entering all fields of life, concerns are raised regarding their potential harmful effects. The simplistic nature of *C. elegans* has helped the study of the effects of CuNPs on reproduction, feeding behavior, and average development.

6 Traditional Management of Nematodes

Nematodes are challenging to control. The commonly used methods are preventive and reduction of the number of nematodes from the upper layer of soil only. Hence, these methods are not very effective beyond one year. The methods include sanitation, choice of plant varieties, fallowing, crop rotation, and soil solarization. Nematicides like fosthiazate are used to control nematodes [33, 34]. Fosthiazate is an organic phosphonate insecticide, an inhibitor of acetylcholinesterase. Fumigant nematicides like methyl iodide or hydrogen cyanide gas can reduce nematodes more effectively. However, they have been banned due to environmental toxicity. Hence, there is a search for cheap and green approaches to managing nematodes and sustaining agricultural productivity. Recent advances in this field show promise in using CuNPs against phytonematodes.

7 Use of CuNPs in the Management of Nematodes

7.1 Green Approaches to Produce CuNPs

A new approach in synthesizing NPs is using plant extracts containing various phytochemicals as the reducing agents. Some of the phytochemicals can also be used as end-capping agents and dispersants in the process of NPs development simultaneously [35]. These methods reduce the use of toxic and harmful reagents. At present, green synthesis of NPs includes using whole plants or various parts of a plant and microorganisms. These green NPs can be degraded more easily and cost-effective. In a green approach study, Khan et al. showed the nematicidal effect of biosynthesized CuONPs against *Meloidogyne incognita*. Mono-disperse, pure CuONPs spherical beads of 5–15 nm diameter were synthesized from *Jatropha curcas* leaf extract and $CuSO_4$, which could kill J2s and inhibit egg hatching at 200 ppm concentration at 24 h of exposure. The CuONPs were found to reduce the root infection caused by *M. incognita*, as evidenced by reduced root galls of chickpeas and enhanced plant growth [36]. The secondary metabolites, e.g., polyphenols, saponins, coumarins, phenols, alkaloids, present in the leaf extract of *J. curcas*, help in the stabilization of CuONPs through the reduction of Cu^{+2} to Cu^{+0}.

Another green approach to produce CuONPs by Khan et al. was from the plant *Spirodela polyrrhiza* (greater duckweed) and cupric sulfate ($CuSO_4$). CuONPs thus produced were evenly distributed and spherical with a diameter in the 100 nm range. These nanoparticles were found to be antibacterial, antifungal, nematicidal, and effective against plant pests and pathogens. The antinematicidal effect increases with exposure time and concentration of the nanoparticles. CuONPs were found to prevent nematode egg hatching in time and concentration-dependent way [37] (Table 1).

Ahmed et al. compared the effects of green synthesized (GBCuNPs) and chemically synthesized (CCuNPs) CuNPs against *M. incognita*. The size of the CCuNPs varies with an average of about 126 nm, whereas the NP size of GBCuNPs was an average of about 85 nm. GBCuNPs were prepared with the medicinal herb *Haplophyllum tuberculatum* and cupric nitrate, $Cu(NO_3)_2{\cdot}3H_2O$. The whole plant extract was used for this purpose. They compared the effects of CCuNPs, GBCuNPs,

Table 1 Green biosynthesized copper nanoparticles used against nematodes

Precursor	Reducing agent (plant/microorganism)	Product description	Key references
$CuSO_4$	*Jatropha curcas*	CuONP ~ 5–15 nm diameter	[36]
$CuSO_4$	*Spirodela polyrrhiza*	CuONPs ~ 100 nm diameter	[37]
–	*Tridax procumbens*	15–45 nm, with ~ particle size of 23 nm	[39]
–	*Bacillus subtilis*	~ 23 nm	[40]
$Cu(NO_3)_2{\cdot}3H_2O$	*Haplophyllum tuberculatum*	~ 85 nm	[38]
$CuSO_4$	*Orobanche aegyptiaca*	> 50 nm	[41]
$Cu(NO_3)_2$, $3H_2O$	Walnut extract	–	[42]

and a positive control chemical nematicide, Nemaprop at three different concentrations (100, 150, and 200 ppm) *in vitro*, against J2 of *M. incognita.* Results reveal that CCuNPs showed the highest mortality of 73.4%, followed by 63.59% of GBCuNPs. The lowest mortality of 6.14% was seen with the chemical nematicide. A concentration-dependent study showed a 90% reduction in 100 ppm of CCuNPs, while a 60.17–70.90% reduction was observed with 150 and 200 ppm of CCuNPs and GBCuNPs [38].

Aqueous extract of *Tridax procumbens*, the perennial herb with copper, iron, and zinc salts, was used to produce respective NPs. In field trials, NPs with 2.98–100 nm particle sizes were formed with potent nematicidal activities against *M. incognita* at 100 ppm on cabbage plants. Cabbage plants treated with *T. procumbens* CuNPs yielded higher than other NPs and normal control cabbage plants. Promising results with CuNPs against *M. incognita* management can replace synthetic nematicides for sustainable cabbage production in the future [39].

It has been found that microorganisms can produce green CuNPs with fungicidal and nematocidal activity. In a study, *Bacillus subtilis* subsp. *subtilis* has been shown to reduce Cu by synthesizing BCuNPs of ~ 23 nm. Application of different concentrations (100–250 ppm) of BCuNPs significantly ($P < 0.05$) reduced the percentage of egg hatching and the number of J2s of *M. incognita.* BCuNPs were most effective on *M. incognita* at 250 ppm after 96 h of application. Also, the mixing of BCuNPs with a known nematicide showed a synergistic effect and was more effective against *M. incognita.* All the parameters used for the measurement of nematode infection, e.g., number of galls, root gall index, egg mass index of *M. incognita* on infected cucumber, and the population density of *M. incognita* J2s reduced significantly [40].

The nematicidal properties of green CuNPs synthesized from the stem extract of the plant *Orobanche aegyptiaca* were checked on *M. incognita* J2. Various concentrations of 50–800 μg/mL of CuNPs were checked for 1, 2, 4, 8, and 16 h. It was found that the percentage mortality of nematodes was directly proportional to the concentration and exposure time of NPs. All the concentrations of CuNPs caused mortality of J2 of *M. incognita* up to 91.5% at 16 h. The highest concentration used in the study, 800 μg/mL, showed the highest mortality of J2 and the highest inhibition of egg hatching. Similar nematicidal and ovicidal properties of other NPs, like ZnO NPs, Ag, Au, and SiO_2 NPs, have been reported [41].

Extracts from walnuts were also found to reduce copper and produce CuONPs [42]. They emphasized the required concentration and time of application of NPs that might have the best results on the growth of the cowpea plant. They concluded that applying 100 ppm CuONPs before *M. incognita* inoculation to cowpea plants exhibited healthier growth parameters along with a reduction in nematode infection compared to control plants. A higher concentration of 200 ppm was more effective at removing the nematode but was harmful to the cowpea plants [42].

7.2 Chemically Synthesized CuNPs on Nematodes

CuNPs have been known as antimicrobials and nematicides for many years. In a study, the researchers showed the nematicidal effects of both copper/iron-based NPs against *Meloidogyne* spp. and their effects on the growth of the nematode-infested plants. The nematicidal activity of PEGylated Cu and CuFe/$CuFeO_2$ NPs was dose-dependent and paralysis of *M. incognita* and *M. javanica* was observed after 24 h. Among the three tested compositions, CuFeNPs were found to be most effective, causing 100% paralysis to J2 of both *Meloidogyne* species at the lowest dose after 72 h. The half maximal effective concentrations (EC50) of Cu, CuFe, and $CuFeO_2$ NPs were 75, 18, and 73 μg/mL respectively for J2 of *M. incognita* after 24 h immersion in the solutions. Similar effects were observed after 48 and 72 h of treatment also. The results established the highest nematicidal activity of CuFeNPs. The EC50 values were 65, 15, and 75 μg/mL for Cu, CuFe, and $CuFeO_2$ NPs against *M. javanica* J2 after 24 h. The results were compared with CuSO4 and fosthiazate, for which the estimated EC_{50} values were 32.9 and 0.79 μg/mL, or 24.5 and 0.89 μg/mL for *M. incognita* or *M. javanica* J2, respectively, after 24 h [43].

The effects of CuNPs were checked on the mortality of *Steinernema feltiae, the* entomopathogenic nematodes (EPNs) present in the biopesticide Owinema biopreparation. The nematodes' mortality rate was positively correlated with the concentrations and exposure time to CuNPs. The highest and lowest percentages of mortality of *Steinernema feltiae* were 48% and 10% at 5 ppm and 0.5 ppm, respectively, for 5 days [44].

CuNPs with polyethylene glycol and Cu-doped ZnO NPs with diethylene glycol with the average size of 12–21 nm were found to be antifungal and antiparalytic against the nematode *Meloidogyne javanica* in a dose-dependent manner. Cu-doped ZnO NPs were more effective against *M. javanica* with an EC_{50} value of 2.60 μg/mL than the EC_{50} value of 25 μg/mL of Cu NPs. The activity of the NPs was checked by spraying them on *Lactuca sativa* plants and inoculated with *M. javanica.* No cytotoxicity was observed in lettuce. The reduction in galling and population of *M. javanica* ranged from 39.32% to 32.29%, which was similar to chemical control

Table 2 Chemically synthesized copper nanoparticles used against nematodes

Precursor	Reducing agent	Product description	Key references
$Cu(NO_3)_2 \cdot 3\ H_2O$	–	CuNPs ~ 126 nm	[38]
$Cu(NO_3)_2 \cdot 3H_2O$ $Fe(NO_3)3 \cdot 9H_2O$	$N_2H_4 \cdot H_2O$	CuNPs, CuFeNPs and $CuFeO_2$NPs ~ 40 nm	[43]
$CuSO_4$, $5H_2O$	Ascorbic acid	CuNPs ~ 100 nm	[46]
$Cu(NO_3)_2 \cdot 3H_2O$, $Zn(NO_3)_2 \cdot 4H_2O$	PEG 8000 $N_2H_4 \cdot H_2O$	Cu-doped ZnO and CuNPs ~ 12–21 nm	[45]

application. The treatment of lettuce plants with Cu-doped ZnO NPs increased photosynthetic value in leaves at 4.60 and 6.66 μmol CO_2 $^{-2}$ s^{-1} when inoculated with *S. sclerotiorum* and *M. javanica*, respectively. The value was 3.00 μmol CO_2 $^{-2}$ s^{-1} in the control set. Phenolic compounds and antioxidant capacity increased in lettuce inoculated with *M. javanica* and treated with Cu-doped ZnO and CuNPs, respectively [45].

Mohammed et al. examined the effects of CuNPs at different concentrations on J2 of *M. incognita* and compared the effects with AgNPs (Table 2). They found a positive linear nematicidal effect of CuNPs against J2 *M. incognita.* CuNPs exhibited 100% mortality of J2 at 200 ppm concentration compared to AgNPs (52%) after 3 days of indirect exposure in soil. CuNPs at a concentration of 0.2 g/L were sufficient to inactivate all nematodes completely. The multifactorial toxic effect of CuNPs might be responsible for its effectivity [46].

8 Studies on *C. elegans* Reveal the Toxicological Effects of CuNPs

In an approach to study the toxic effects of CuNPs on development and neuronal behavior, Mashock et al. examined and compared the toxicological effects of CuONPs and $CuSO_4$ on *C. elegans* [47]. They studied three wild *C. elegans* isolates with one laboratory strain to determine the influence of different genotypes on the physiological response to Cu exposure. Reduced body length, change in feeding behavior, and reduced reproduction rate were observed in *C. elegans* on exposure to Cu, with CuONPs showing prominent effects compared to $CuSO_4$. Significant neurodegeneration was observed after treatment with Cu and CuNPs. The normal neuronal morphology of a transgenic *C. elegans* strain with dopaminergic neurons expressing GFP was changed when exposed to CuONPs and $CuSO_4$ in a dose-dependent manner. Neurodegeneration was more prominent when exposed to CuONPs than $CuSO_4$ with equal Cu content. In addition, the effects of exposure to Cu were much less in mutant strains containing knockouts of either *smf-1* or *smf-2* (deletion of a divalent metal transporter gene) compared to the BY250 wild-type strain. No degeneration was observed in the untreated animals of transgenic strain.

Adding live bacteria to the solution of NPs has been shown to increase their effectiveness. It is known that the suspension of CuONPs contains significantly fewer copper ions compared to the same concentration of $CuSO_4$ solution (twofold more efficacy) ($p < 0.04$). However, the toxicological effects of CuONPs were greater than those of $CuSO_4$ as greater sensitivity of *C. elegans* to CuONPs exposure was observed. The addition of live bacteria in NP solution increases the adsorption of free Cu ions on their surface. During the consumption of bacteria, NP concentration increases inside the nematodes [48]. Ingestion of CuONPs, along with food, is the entry point of the NPs in *C. elegans*, as the cuticles on the nematode are impermeable to most of the chemicals. Accumulation of CuONPs in the intestine leads to limited food uptake by *C. elegans,* thus reducing its motility [49].

In a similar experiment to assess the accumulation and toxicity of Cu in *C. elegans*, four different conditions or pathways were chosen. In one, Cu was in the food; in the second one, Cu was in the water; next, water and food both had Cu in them, and in the fourth, there was clean water but food with Cu. In the case of Cu in food, *C. elegans* had maximum exposure to Cu, which is evident from Cu accumulation, less growth status, feeding ability, and upregulated antioxidant enzymes. The exposure pathway without food had a minor effect on the growth and antioxidant activities in *C. elegans* and had the best growth statuses. Hence, Cu-containing food was the primary exposure pathway; the other pathways had different responses on *C. elegans* [50].

The bioaccumulation and toxicological effects of Cu^{2+} and CuNPs were studied on *C. elegans* by Gao et al. by using the scanning technique of microbeam synchrotron radiation X-ray fluorescence (μ-SRXRF). The non-destructive and multi-elemental μ-SRXRF mapping indicated an overall increase in Cu and K concentrations and uneven distribution of Cu in *C. elegans.* Excess Cu was found in the head and upper body compared to the normal control when treated with CuNPs. Higher amounts of Cu were detected in the intestine and excretory cells when exposed to Cu^{2+}. The total amounts of Cu detected were 4.10 ± 0.54, 12.32 ± 0.49, and 5.22 ± 0.63 μg g^{-1} dry weight of *C. elegans* in normal control, Cu^{2+} and CuNP groups respectively [51].

9 Conclusion

Nanotechnology has already created a multibillion-dollar world market. Nanomaterials/nanoparticles have promising uses in fundamental and essential sectors, such as energy, agriculture, communications, pollution control, new and improved materials and products, and medical and biomedical applications. Metal NPs, e.g., AgNPs and CuNPs, are of immense importance in agriculture as nematicides. CuNPs are highly effective against nematodes, the major cause of crop destruction. The increasing popularity of CuNPs is because of their low cost and high effectivity compared to other metal NPs. The higher toxicity of CuNPs is due to their smaller size, high penetration power, and increased ROS generation. Green biosynthesized CuNPs are even better, as the reducing agents used here are plant materials or microorganisms, which are easily degradable and less toxic to the environment. However, we should be careful when using NPs, as they enter the environment during various human activities and lead to ecotoxicity influenced by abiotic factors like pH, salinity, organic factors, etc.

Competing Interests The authors declare that they have no competing interests.

Declaration of Generative AI and AI-Assisted Technologies in the Writing Process During the preparation of this work the book chapter author(s) have not taken any assistance from AI, and the book chapter author(s) reviewed and edited the content as needed and take(s) full responsibility for the content of the publication. All authors read and approved the final version of the manuscript.

Abbreviations

CCuNPs	Chemically synthesized copper nanoparticle
Cu	Copper
CuONPs	Copper oxide nanoparticles
CuNPs	Copper nanoparticles
$Cu(NO_3)_2 \cdot 3H_2O$	Copper nitrate trihydrate
$CuSO_4$, $5H_2O$	Copper sulfate penta hydrate
EPA	Environmental protection agency
$Fe(NO_3)3 \cdot 9H_2O$	Ferric nitrate pentahydrate
GBCuNPs	Green biosynthesized copper nanoparticle
J1	Juvenile first stage
J2	Juvenile second stage
$N_2H_4 \cdot H_2O$	Hydrazine monohydrate
NPs	Nanoparticles
PEG	Polyethylene glycol
RKN	Root knot nematode
ROS	Reactive oxygen species
μ-SRXRF	Scanning technique of microbeam synchrotron radiation X-ray fluorescence
$Zn(NO_3)_2 \cdot 4H_2O$	Zinc nitrate tetrahydrate

References

1. Bhatia S (2016) Nanoparticles types, classification, characterization, fabrication methods and drug delivery applications. In: Natural polymer drug delivery systems. Springer, Cham. https://doi.org/10.1007/978-3-319-41129-3_2

2. Tiwari JN, Tiwari RN, Kim KS (2012) Zero-dimensional, one-dimensional, two-dimensional, and three-dimensional nanostructured materials for advanced electrochemical energy devices. Prog Mater Sci 57:724–803. https://doi.org/10.1016/j.pmatsci.2011.08.003
3. Zhang J, Saltzman M (2013) Engineering biodegradable nanoparticles for drug and gene delivery. Chem Eng Prog 109(3):25–30
4. Prashant KJ, Ivan HS (2007) Au NPs target cancer. Nanotoday 2:19–29
5. Ibupoto Z, Khun K, Beni V, Liu X, Willander M (2013) Synthesis of novel CuO nanosheets and their non-enzymatic glucose sensing applications. Sensors 13(6):7926–7938. https://doi.org/10.3390/s130607926.PMC3715261
6. Dreaden EC, Austin LA, Mackey MA, El-Sayed MA (2012) Size matters: gold nanoparticles in targeted cancer drug delivery. Ther Deliv 3(4):457–478. https://doi.org/10.4155/tde.12.21
7. Bahadar H, Maqbool F, Niaz K, Abdollahi M (2016) Toxicity of nanoparticles and an overview of current experimental models. Iran Biomed J 20(1):1–11. https://doi.org/10.7508/ibj.2016.01.001
8. Ferreira AJ, Cemlyn-Jones J, Cordeiro CR (2013) Nanoparticles, nanotechnology and pulmonary nanotoxicology. Rev Port Pneumol 19:28–37
9. Navarro E, Piccapietra F, Wagner B, Marconi F, Kaegi R, Odzak N, Sigg L, Behra R (2008) Toxicity of silver nanoparticles to *Chlamydomonas reinhardtii*. Environ Sci Technol 42:8959–8964
10. Azhar W, Khan AR, Salam A et al (2023) Ethylene accelerates copper oxide nanoparticle-induced toxicity at physiological, biochemical, and ultrastructural levels in rice seedlings. Environ Sci Pollut Res 30:26137–26149. https://doi.org/10.1007/s11356-022-23915-8
11. Khan FA (2011) Biotechnology fundamentals. CRC Press, Boca Raton
12. Dhas NA, Raj CP, Gedanken A (1998) Synthesis, characterization, and properties of metallic copper nanoparticles. Chem Mater 10(5):1446–1452. https://doi.org/10.1021/cm9708269
13. Bogdanović U, Lazić V, Vodnik V, Budimir M, Marković Z, Dimitrijević S (2014) Copper nanoparticles with high antimicrobial activity. Mater Lett 128:75–78
14. El-Samad LM, El-Gendy AH, Abdel-Moneim AM, Saeed El-Ashram S, Maria AM (2022) CuO NPs-induced damage to testes and deregulation of the antioxidant system in wild terrestrial organism *Blaps sulcata* (Coleoptera: Tenebrionidae). Environ Nanotechnol Monit Manag 18:100751
15. Handy RD, Kammer R, Lead JR, Hassellöv M, Owen R, Crane M (2008) The ecotoxicology and chemistry of manufactured nanoparticles. Ecotoxicology 17:287–314
16. Xia T, Kovochich M, Liong M, Mädler L, Gilbert B, Shi H, Yeh JI, Zink JI, Nel AE (2008) Comparison of the mechanism of toxicity of zinc oxide and cerium oxide nanoparticles based on dissolution and oxidative stress properties. ACS Nano 2:2121–2134
17. Johnson FH, Carver CM, Harryman WK (1942) Luminous bacterial auxanograms in relation to heavy metals and narcotics, Self-photographed in color. J Bacteriol 44(6):703–715. https://doi.org/10.1128/jb.44.6.703-715.1942
18. Michels HT, Noyce JO, Keevil CW (2009) Effects of temperature and humidity on the efficacy of methicillin-resistant Staphylococcus aureus challenged antimicrobial materials containing silver and copper. Lett Appl Microbiol 49:191–195
19. Weaver L, Michels HT, Keevil CW (2010) Potential for preventing spread of fungi in air-conditioning systems constructed using copper instead of aluminium. Lett Appl Microbiol 50(1):18–23. https://doi.org/10.1111/j.1472-765X.2009.02753.x
20. Vincent M, Duval RE, Hartemann P, Engels-Deutsch M (2018) Contact killing and antimicrobial properties of copper. J Appl Microbiol. 24(5):1032–1046. https://doi.org/10.1111/jam.13681. Epub 2018 Feb 2. PMID: 29280540
21. Budiyanti DS, Moeller ME, Thit A (2022) Influence of copper treatment on bioaccumulation, survival, behavior, and fecundity in the fruit fly *Drosophila melanogaster*: toxicity of copper oxide nanoparticles differ from dissolved copper. Environ Toxicol Pharmacol 92:103852
22. El Bialy BE, Hamouda RA, Abd Eldaim MA, El Ballal SS, Heikal HS, Khalifa HK, Hozzein WN (2020) Comparative toxicological effects of biologically and chemically synthesized copper oxide nanoparticles on mice. Int J Nanomedicine 28(15):3827–3842. https://doi.org/10.2147/IJN.S241922

23. Ahmed M, Hisham A, Alhadlaq HA, Khan MAM, Karuppiah P, Al-Dhabi NA (2014) Synthesis, characterization, and antimicrobial activity of copper oxide nanoparticles. J Nanomater 637858. https://doi.org/10.1155/2014/637858
24. Keller AA, McFerran S, Lazareva A, Suh S (2013) Global life cycle releases of engineered nanomaterials. J Nanoparticle Res 2013(15):1692. https://doi.org/10.1007/s11051-013-1692-4
25. Lamberti F, Taylor CE (eds) (1979) Root-knot nematodes (*Meloidogyne species*): systematics, biology and control. Academic Press, New York City, ISBN 0-12-434850-5
26. Feyisa B (2021) Survey and identification of plant parasitic nematodes on faba bean crop in ethiopia. J Plant Pathol Microbiol 12:261
27. Pline D, Dusenbery J (1988) Extremely sensitive thermotaxis of the nematode Meloidogyne incognita. Nematology 20:605–608
28. Diez JA, Dusenbery DB (1989) Preferred temperature of *Meloidogyne incognita*. J Nematol 21:99–104
29. Perry RN, Starr JL (2009) Root-knot nematodes. Centre for Agriculture and Bioscience International (CABI), London. ISBN 9781845934927
30. Brenner S (1973) The genetics of behavior. Br Med Bull 29:269–271
31. Brenner S, Wood WB (eds) (1988) The nematode *Caenorhabditis elegans*. Cold Spring Harbor Laboratory, Cold Spring Harbor, New York
32. Sulston JE, Schierenberg E, White JG, Thomson JN (1983) The embryonic cell lineage of the nematode Caenorhabditis elegans. Dev Biol 100(1):64–119. https://doi.org/10.1016/0012-1606(83)90201-4
33. Wang H, Yan W, Luo J, Zeng X, Zhao Z, Dou X, Fu M (2023). Fosthiazate inhibits root-knot disease and alters rhizosphere microbiome of Cucumis melo var. saccharinus. Front Microbiol 13–2022. https://doi.org/10.3389/fmicb.2022.1084010/full
34. Wu HY, Luo M, Zhang LY, Zhou XB (2019) Nematicidal activity of fosthiazate against soybean cyst nematode *Heterodera glycines*. J Nematol 51:1–9. https://doi.org/10.21307/jofnem-2019-021.PMID:31088033;PMCID:PMC6929653
35. Devi HS, Boda MA, Shah MA, Parveen S, Wani AH (2019) Green synthesis of iron oxide nanoparticles using Platanus orientalis leaf extract for antifungal activity. Green Process. Synthesis 8(1):38–45
36. Khan A, Bani Mfarrej MF, Danish M, Shariq M, Khan MF, Ansari M, Hashem M, Alamri S, Ahmad F (2022) Synthesized copper oxide nanoparticles *via the* green route act as antagonists to pathogenic root-knot nematode, *Meloidogyne incognita*. Green Chem Lett Rev 15:3:491–507. https://doi.org/10.1080/17518253.2022.2096416
37. Khan M, Khan AA, Parveen A, Min K, Yadav VK, Khan AU, Alam M (2023) Mitigating the growth of plant pathogenic bacterium, fungi, and nematode by using plant mediated synthesis of copper oxide nanoparticles (CuO NPs). Green Chem Lett Rev 16(1):2177520. https://doi.org/10.1080/17518253.2023.2177520
38. Soliman AMM, Abdallah EAM, Hafez EE, Kadous EA, Kassem FA (2022) Nematicidal activity of chemical and green biosynthesis of copper nanoparticles against root-knot nematode. Meloidogyne Incogn Alex Sci Exch J 43(4):583–591
39. Fabiyi O, Lateef A, Gueguim-Kana EB, Beukes LS, Matyumza N, Bello T, Olatunji G (2023) Characterization and nematicidal potential of copper, iron and zinc nanoparticles synthesized from *Tridax procumbens* L. extract on *Meloidogyne incognita* infected cabbage plants. Eur J Plant Pathol. https://doi.org/10.1007/s10658-023-02804-x
40. El-Ashry RM, Nader MM, Shami A, Alduwish MA, Ahmed AE, Alamoud SA, Allohibi A, Alqahtani FS, Alghamdi AM, Ahmed AI, Hassan MAA, Awad SE, El-Tarabily KA (2023) Ecofriendly synthesis and nematicidal application of copper nanoparticles fabricated from *Bacillus subtilis* AM18, against root-knot nematode of cucumber. European J Plant Pathol. https://doi.org/10.1007/s10658-023-02727-7
41. Akhter G, Khan A, Ali SG, Khan TA, Siddiqi KS, Khan HM (2020) Antibacterial and nematicidal properties of biosynthesized Cu nanoparticles using extract of holoparasitic plant. SN Appl Sci 2:1268

42. Tauseef A, Hisamuddina GJ, Rehman A, Uddin I (2021) Differential response of cowpea towards the CuO nanoparticles under *Meloidogyne incognita* stress. S Afr J Bot 139:175–182
43. Ch G, Ntalli N, Menkissoglu-Spiroudi U, Dendrinou-Samara C (2019) Essential metal-based nanoparticles (copper/iron nps) as potent nematicidal agents against *Meloidogyne* spp. J Nanotechnol Res 1(2):044–058. https://doi.org/10.26502/jnr.2688-8521004
44. Kucharska K, Pezowicz E, Tumialis D, Kucharski D, Zajdel B (2014) Nanoparticles of copper and entomopathogenic nematodesSteinernema feltiae (Filipjev, 1934) in reducing the number of thelesser mealworm beetle Alphitobius diaperinus (Panzer, 1797). Ann Wars Univ Life Sci SGGWAnimal Science No 53:29–35
45. Tryfon P, Nathalie N. Kamou, Ntalli N, Mourdikoudis S, Katerina Karamanoli K, Karfaridis D, Menkissoglu-Spiroudi U, Samara CD (2022) Coated Cu-doped ZnO and Cu nanoparticles as control agents against plant pathogenic fungi and nematodes. NanoImpact 28:100430
46. Mohamed EA, Elsharabasy SF, Abdulsamad D (2019) Evaluation of in vitro nematicidal efficiency of copper nanoparticles against root-knot nematode *Meloidogyne Incognita*. South Asian J Parasitol 2(1):1–6, Article no.SAJP.46065
47. Mashock MJ, Zanon T, Kappell AD, Petrella LN, Andersen EC, Hristova KR (2016) Copper oxide nanoparticles impact several toxicological endpoints and cause neurodegeneration in *Caenorhabditis elegans*. PLoS One 2;11(12):e0167613. https://doi.org/10.1371/journal.pone.0167613
48. Ellegaard-Jensen L, Jensen KA, Johansen A (2012) Nano-silver induces dose-response effects on the nematode *Caenorhabditis elegans*. Ecotoxicol Environ Saf 80:216–223. https://doi.org/10.1016/j.ecoenv.2012.03.003
49. Song S, Guo Y, Zhang X, Zhang X, Zhang J, Ma E (2014) Changes to cuticle surface ultrastructure and some biological functions in the nematode *Caenorhabditis elegans* exposed to excessive copper. Arch Environ Contam Toxicol 66:390–399. https://doi.org/10.1007/s00244-013-9991-4
50. Yu ZY, Zhang J, Yin DQ (2012) Toxic and recovery effects of copper on *Caenorhabditis elegans* by various food-borne and water-borne pathways. Chemosphere 87(11):1361–1367. https://doi.org/10.1016/j.chemosphere.2012.02.029
51. Gao Y, Liu N, Chen C, Luo Y, Li Y, Zhang Z, Zhao Y, Zhao B, Iida A, Zhifang Chai Z (2008) Mapping technique for biodistribution of elements in a model organism, *Caenorhabditis elegans*, after exposure to copper nanoparticles with microbeam synchrotron radiation X-ray fluorescence. J Anal Atomic Spectrom 23(8). https://doi.org/10.1039/B802338G

Effect of Copper Nanomaterials in Soil Microbial Communities

Versha Dixit, Pranjali Bajrang Chole, Kundan Kumar Chaubey, Saswati Barman, Shreya Pandey, Gaurav Bhardwaj, Anish Kumar Pal, Ashwani Kumar Sanghi, and B. T. Manjunath

Abstract A growing number of agricultural applications focus on copper nanoparticles (Cu NPs) because of their multifunctional qualities and many unique characteristics. This chapter covers the application of copper nanoparticles (Cu NPs) in the agrochemicals field for releasing pesticides and fertilizers and providing efficient delivery. The use of agrochemicals at extreme levels and for a long time may harm the soil system. Additionally, they have strong antimicrobial properties that inhibit the activities of soil microorganisms so that they may be harmful to the soil's health and the microorganisms crucial to the operation of both terrestrial and agroecosystems. Therefore, it is essential to understand the effects, use, and mechanism of action of copper nanoparticles (Cu NPs) on soil processes, soil microorganisms, plants, and their distinctive nano-specific properties. This chapter reviews the most recent research on the possible effects of copper nanoparticles (Cu NPs) on the soil–plant ecosystem in the agriculture field. It gives readers a comprehensive grasp of the copper nanoparticles in the soil–plant ecosystem. This chapter explores the extensive use of copper nanoparticles in agriculture, their potential interactions with different soil constituents, and their effects on variables. This chapter also explores how plants absorb, move and distribute them across various exposure and extreme conditions. This chapter summarizes that several variables such as pH, types of soil and organic matter concentration control and affect the solubility, bioavailability and biotransformation of copper nanoparticles (Cu NPs) in the soil–plant ecosystem. Other factors,

V. Dixit · P. B. Chole · B. T. Manjunath
Department of life sciences, CHRIST (Deemed to be University), Bangalore, India

K. K. Chaubey (✉) · S. Barman · S. Pandey · G. Bhardwaj
Department of Biotechnology, School of Basic and Applied Sciences, Sanskriti University, Mathura, India
e-mail: kundan2006chaubey@gmail.com

A. K. Pal
Division of Research and Innovation, School of Applied and Life Sciences, Uttaranchal University, Dehradun, India

A. K. Sanghi
School of Allied and Health Sciences, MVN University, Palwal, Haryana, India

R. K. Bachheti et al. (eds.), *Copper-Based Nanomaterial for Agricultural Practices*, Smart Nanomaterials Technology, https://doi.org/10.1007/978-981-96-8610-0_12

such as contaminants, environmental conditions, and time intervals, may also affect their soil toxicity and biotransformation.

Keywords Copper nanoparticles · Agroecosystem · Engineered nanoparticles · Soil microorganism · Cu NP-based nanocomposites

1 Introduction

A worldwide challenge is raising agricultural output economically without endangering the environment. Despite recent significant advancements in agriculture and healthcare, food and healthcare insecurity remains a problem in many developing nations. Scientists are working to create antimicrobial agents such as antimicrobial peptides, cationic polymers, and metal nanoparticles to combat drug-resistant microorganisms by utilizing the most recent advancements in nanobiotechnology [24, 37, 48]. Nanotechnology has recently been used to increase crop yield in agriculture, although its uses in post-harvest food product storage have been more widespread. It has been noticed that the nanotechnology is being used to reduce the effects of both biotic and abiotic stresses to increase crop yield and plant growth [2–4, 12]. The vital micronutrient copper, incorporated into numerous proteins and enzymes, is crucial to the health and nourishment of plants. Researchers are interested in employing copper nanoparticles as a crucial component in future nanodevices because of copper's exceptional characteristics, such as suitable catalytic activities, good electrical conductivity, and surface-enhancing Raman scattering activity [7, 18]. With the current issues in food security and pollution, advancements based on nanotechnology provide a critical and promising chance to revitalize the agriculture sector sustainably. Copper nanoparticles (Cu NPs) or engineered nanoparticles (ENPs) are synthesized by modification at various nanotechnology levels, such as macromolecular, atomic, and molecular levels. By using the microorganisms, the delivery of fertilizers and pesticides via nanoparticles (NPs) reduces the ecological impact of agricultural practices [5].

Nanoparticles such as metal oxide engineered nanoparticles (ME NPs) and copper oxide nanoparticles (CuO NPs) are broadly applied as consumer goods, energy transmission, sensitive gas sensors, and antibacterial agents in chemical degradation [14, 49]. Due to their antibacterial characteristics, these Copper oxide nanoparticles (CuO NPs) increase crop production and the efficiency of Cu delivery to soil–plant ecosystem. They are also broadly used as insecticides, fungicides, and nanoscale fertilizers. Antibacterial properties against the pathogens present in soil have also been proven by Copper oxide nanoparticles (CuO NPs) [5]. The explosive use of Copper oxide nanoparticles (CuO NPs) is done in commercial products such as agrochemicals, antimicrobial products, catalyzers, compounds, paints, sensors, and semiconducting products, So there is a continuous increase in the production of copper nanoparticles (Cu NPs) in both aquatic and terrestrial ecosystems [19]. The newly discovered pollutants can enter the soil by continuously using biosolid products produced from wastewater treatment plants or by directly using copper-containing nano pesticides or fertilizers [16]. Copper nanoparticles have antibacterial qualities and strong reactivity, which stimulates worries regarding the possible effects on the microbial activities of soil that maintain soil fertility in agroecosystems [43]. Soil has a significant role in both the exchange of energy and substances in the ecosystem. So, it is necessary to maintain the safety and improve the quality of the soil for the purpose of equilibrium and stabilizing the whole ecosystem. The soil's biological, chemical, and physical characteristics, such as chemical composition, pH, redox potential, texture, structure, temperature, humidity, and organic content, determine and maintain the soil's utility and quality [42].

Copper Oxide Nanoparticles (CuO NP) bioavailability and toxicity are probably affected by soil characteristics such as ionic strength, organic matter, pH, texture. As a result, findings from one kind of soil should not be applied to other soil types [8]. For instance, the properties of the soil might affect how Copper Oxide Nanoparticles (CuO NP) evolve through dissolution processes [19]. The Cu ion at high concentrations could be poisonous, but at low concentrations, it becomes a necessary element for soil microorganisms for biological growth [1].

2 Plant, Soil, and Copper-Based Nanoparticles: Use and Mechanism of Action

Cu-based NPs have antibacterial qualities, are small in size, and have an ion-releasing capacity. Copper Nanoparticles (Cu NP) can work as an antimicrobial agent to safeguard crops from pests and illnesses because of their diverse action against microorganisms; cu-based NPs having tiny sizes are more accessible for plants to absorb and have improved biological efficacy, which reduces overuse and runoff risk (Fig. 1). Ultimately, this may lessen the detrimental effects that agriculture has dependent on the agroecosystem, especially the soil system [13]. Copper nanoparticles (Cu NP) can be produced using techniques such as green synthesis and chemical techniques. Cu-based NPs are considered safer than chemical production because of their low impact on the environment and the green synthesis, which uses plants and microorganisms. Several research organizations have produced copper Nanoparticles (Cu NP) and have also examined how they work against the microorganisms, bacteria, and fungi that cause plant diseases [5]. For example, Cu-based NPs have been produced by using a metal-vapor synthesis (MVS) method [39] and assessed the antimicrobial activity by impregnating chitosan in Copper Nanoparticles (Cu NP) solution. These Copper Nanoparticles (Cu NP) composites inhibited the *Rhizoctoniasolani* AG-4 and *Sclerotiumrolfsii* plant pathogenic fungi. Cytoplasmic coagulation and hyphae eradication were used to inhibit fungal growth. Cu-chitosan NPs produced using the ionic gelation technique involve cross-linking chitosan and tripolyphosphate (TPP) ions [40]. They examined how well they could promote tomato plant growth and fight against fungi like *Alternariasolani* and *Fusariumoxysporum.* Using a chemical

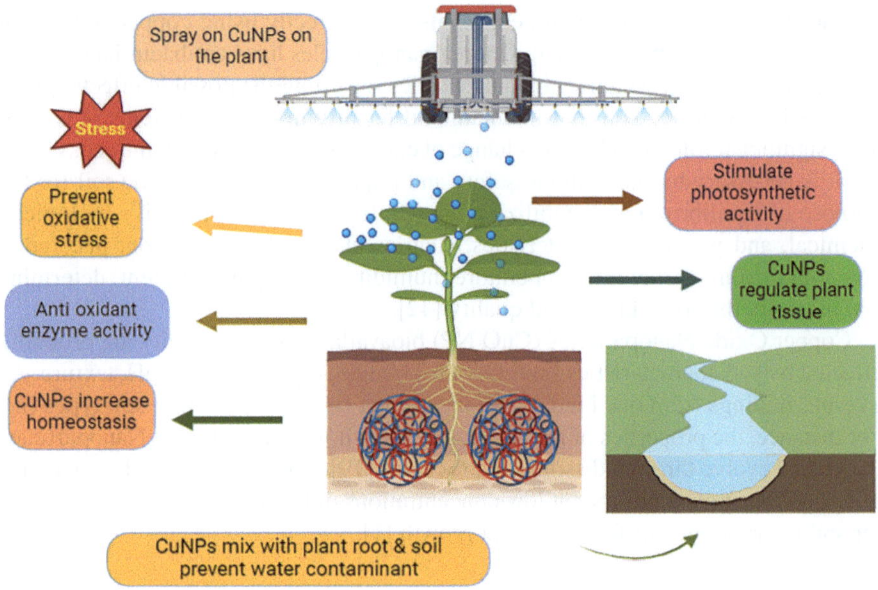

Fig. 1 Copper nanoparticles of plant increase photosynthetic rate, prevent oxidative damage, and prevent water contaminant

reduction technique, Copper Nanoparticles (Cu NP) coated with 3–10 nm size Cetyl Trimethyl Ammonium Bromide (CTAB) showed high antifungal efficacy against four fungi. The size of the zone of inhibition for *Fusariumoxysporum*, *Phomadestructive*, *Curvularialunata,* and *Alternariaalternata* is in the order 18, 21, 22, and 24 mm, respectively [17]. To synthesize copper Nanoparticles (Cu NP), ascorbic acid and sodium citrate are used as active reducing agents, dispersion agents, and protective agents. Copper Nanoparticles (Cu NP) had fungicidal activities against *Fusariumsolani* sp., *Fusariumoxysporum*at sp. And *Neofusicoccum* sp. Damage to cell membranes in fungi is induced by the large production of intracellular reactive oxygen species (ROS) production in the mycelium, which accounted for their antifungal activity [32].

3 Effect of Copper Nanoparticles (Cu NP) on Soil and Soil Microorganisms

Nanoparticles such as metal oxide-engineered nanoparticles (ME NPs) are spread into the soil by sewage sludge in agricultural fields. Metal-engineered nanoparticles (ME NPs) are disposed of into wastewater treatment methods through waste disposal pathways and are mixed up in agriculture sludge. Agriculture sludge are utilizing these Metal-engineered nanoparticles (ME NPs) as biosolids that carry specific unidentified concerns, including the possibility of biomagnification or food chain transmission [5].

Furthermore, the situation may worsen if these Metal-engineered nanoparticles (ME NPs) amendments containing agriculture sludge are repeatedly applied to agricultural soils [6]. Furthermore, the intentional and direct introduction of ENPs into the soil would result from using nanoparticles containing agrochemicals. This is especially true for copper nanoparticles (Cu NP), primarily used in agrochemicals to control pests and plant diseases. Using these Copper Nanoparticles (Cu NP) or agrochemicals and their products may affect and spread throughout the land ecosystem, especially the soil system, in a new manner. Near about 95% of all copper discharged into the environment will wind up in aquatic silt and soil, where the concentration build limit of the element is 500 μg/L [19]. This soil influx could be explained by a significant amount of Copper Nanoparticles (Cu NP) deposited into biosolids. Copper Nanoparticles (Cu NP) affect the physical elements of soil, such as ion exchange capacity (IEC), pH, redox potential, and organic matter of soil [33]. Changing the enzymes or processes necessary for the soil's biogeochemical cycles may also affect the soil. Finally, they significantly affect the nature and organization of microbial communities in soil and the interactions between plants, soil, and microbes [5].

The growth of soil microorganism is an indicator and parameter of the health of the soil. When \Copper Nanoparticles (Cu NP) are mixed in to the agriculture soil and water, it has been observed that they can harm the growth of soil microorganism and enzyme activity of soil microorganism. In the agriculture soil, copper nanoparticles (10, 100, and 1000 mg/kg) more potently suppressed the activities of various enzymes such as dehydrogenase, acid phosphatases, alkaline phosphatases, and ureases. After a 30-day exposure (1, 10 and 100 mg/kg soil), Copper Nanoparticles (Cu NP) and Cu ions enhanced the activity of several extracellular enzymes such as 1,4-glucosidase, 1,4-N-acetylglucosaminidase, 1,4-xylosidase, acid phosphatase and cellobio hydrolase. Copper nanoparticles (Cu NP) can affect soil microorganisms' growth and enzyme activity. Similarly, exposure to Copper Nanoparticles (1 to 200 mmol CuO/g soil) inhibits the growth of soil microorganisms, and it is essential to note that, however, there are adverse effects on the dose-dependent [35]. Moreover, Copper Nanoparticles (Cu NP) are also associated with other pollutants affecting the growth of agricultural soil microorganisms. When the Copper Nanoparticles (0.05% to 0.15% w / w) and the herbicide atrazine (3 mg/kg) exist at the same time, then these affect the amount of nitrifying bacterial communities, and these changes gradually stay up to 30 days [31].

4 Treatments by Using Copper Nanoparticles (Cu NP) in Soil and Soil Microorganisms

The size and form of copper affects the soil and soil microorganisms. Micro-X-ray fluorescence microscopy and X-ray absorption near edge structure spectroscopy revealed the size-dependent CuO dissolution. More significant increases in Cu made accessible for soils treated with smaller nanoparticles provided evidence. When $CuCl_2$ was added, the precipitation and hydrolysis processes changed the soil to acidic. However, Copper Nanoparticles (Cu NP) applications raised the soil pH in just one hour. Land use largely influenced the microbiological response to Cu treatment; even the Copper Nanoparticles significantly affect the chemical factors such as the pH of the soil and the available Cu of soil. As a result, the reactions that were unique to the exchange emerged [38].

Unmanaged Cu amendment systems affect agricultural soil microorganisms' growth by allowing them to have the maximum respiration (activity) and microbial biomass per unit. In conventional and nonconventional systems, exposure to Cu did not influence the microbial biomass; however, in all soils, CuCl2 at 1000 mg/kg soil reduced respiration. This recommends that the nutrient cycling is driven by microorganisms, which is essential for boosting the availability of nutrients, especially in agriculture farming systems, and may be revealed by even small changes in the amount of microbial biomass. Moreover, historical use often influences microbiological responses to Cu amendment more than the type or pace of Cu treatment. The unique relationship has shown that soil respiration, agricultural microorganism production, and copper treatment have increased each land use. Thus, historical usage of agricultural land must be considered when interpreting the results of Copper Nanoparticles to soil toxicity tests and also other metal toxicity tests [38].

The substantial variation in particle size was attributed to the substantial fraction of massive molecules (more than > 1000 nm) in the main CuO molecules population. Despite significant overlap in particle sizes, the CuO molecule sizes had varying diameters of molecules. All Copper Nanoparticle speciation was established using the X-ray diffraction technique, and the CuO speciation of 16n CuO and 42n CuO was confirmed using bulk X-ray diffraction absorption technique near the edge structure spectroscopy. The diffractograms and XANES spectra showed X-ray diffraction absorption peaks as particle size decreased [20]. This is due to increased signal scattering caused by the nanoparticles' smaller crystalline domain size. The decreasing diffractogram intensities and increased proportion of surface defects per unit mass for copper nanoparticles decrease the degree of crystallinity of CuO. Due to these defects, Copper Nanoparticles may dissolve more efficiently than bulk Copper Nanoparticles [38].

5 Effect of Copper Nanoparticles on Toxicity and Properties of Soil

Soil is an immensely useful receptor of nanoparticles and a rich source of natural nanoparticles, i.e., both primary and secondary nanoparticles. Artificial soil exposure to nanoparticles showed a remarkable impact, but it can accumulate in the soil and resist degradation. Different factors like concentration, enzymatic activity, and soil type are responsible for the effect of nanoparticles on the soil. Adverse effects on the soil were observed on dehydrogenase enzyme activity when using high concentrations of nanoparticles [15]. Nanoparticle toxicity affects the soil properties. Soil properties include texture, organic content, and structure. pH, soil microbial community influence, and pollutant toxicity toward microorganisms [9, 36].

6 Effect of Copper Nanoparticles on Microbiota

Expeditious enlargement use of various metallic nanoparticles in biosensors, catalysts, electronics, environmental remediation, medicine, optics, textiles, and wastewater treatment has expanded their importance on agriculture soil, the environment, and the healthcare system. Excessive use of nanoparticles can increase the toxic effects on different organisms, including fungi, bacteria, yeasts, fish, diatoms, and healthcare systems [25, 30]. Plants and soil microorganisms are common nanoparticle receptors, and the biomass of soil microorganisms is very effective in microbial change of soil [36]. Plants and soil are commonly associated with each other in the 'plant-soil system'. The direct effect of nanoparticles in soil may affect both soil and plant microorganisms, ultimately affecting consumers or the healthcare system [36]. The introduction of nanoparticles at excess levels into the environment may harm beneficial microorganisms like bacteria and fungi. Accumulation of nanoparticles in soil may happen due to excess use of nanoparticles in agriculture (in the form of fertilizers and pesticides), water or soil remediation technologies, and different unintentional releases via water, air, sewage, and sludge, [36, 45, 46]. Different enzymatic activities and soil respiration can help determine nanoparticles' effect on soil microbial activity [44]. Various studies concluded that copper nanoparticles and TiO_2 decreased the enzymatic activity and biomass of microbiota present in soil and drastically affected paddy soil [50]. These results showed that using Copper Nanoparticles changes the microbial community of soil, enzyme activities of soil and biological nitrogen fixation. In the case of copper nanoparticles, copper ions were released and became known as a major cause of destruction to pathogenic bacteria, which is beneficial for the soil [23]. In another study, it was observed that Copper Nanoparticles showed high toxicity due to the formation of blebs, cavities, cellular collapse, cell lysis, and holes in soil bacterial isolates [34]. It is also observed that AgNPs and CuONPs exposure can lead to a decrease in microbial leaf decomposition rate [10].

7 Effect of Copper Nanoparticles in Agriculture

A bordeaux mixture, a combination of calcium hydroxide and copper sulfate, has been used as the antifungal agent from 1882 [5]. It has been studied that Copper Nanoparticles show antimicrobial properties and can be used in agriculture because of their ion-releasing ability (Fig. 2). These Copper Nanoparticles can be used as antimicrobial agents to protect crops against diseases and pests. Similarly, an increased use of CuNPs can affect plant development, physiological changes, and growth [5]. Products based on copper (Cu), like copper oxide nanoparticles (nCuO), are essential in agriculture as fungicides and algaecides. Applying Cu compounds, especially nCuO, to crops and the aerosol drift of these products onto the soil can alter microbial populations and nutrient cycling in controlled and uncontrolled environments. Several researchers observed the effects of land utilization on soil microbial respiration biomass in response to adding nCuO to alluvial soil. Samples of soils were treated with different Copper Nanoparticles with 1000 mg of copper in the form of $CuCl_2$, 16 nm and 42 nm CuO, and bulk CuO (bCuO) larger nanoparticle size. $CuCl_2$ > 16nCuO > 42nCuO > bCuO was the order of copper availability in all soil samples when Cu was added to the soils. All land uses combined, the soils treated with bCuO showed the lowest Cu availability after 70 days [38]. Because of their multitude of promising applications and affordability, Cu-based NPs have gained popularity among metallic NPs. Several experiments were conducted to investigate the effect of nCuO nanoparticles on microbial growth, activity, and diversity. Microbial diversity was also assessed using the community-level physiological profiling (CLPP) method and 16S ribosomal DNA (rDNA) sequencing method, and microbial growth was measured using the heterotrophic plate count (HPC) method. Soil Microbial activity was measured by soil nitrification, organic matter degradation, soil respiration (basal and substrate-induced), and enzyme activity for phosphatase, glucosidase, and dehydrogenase activity of soil. A readily soluble positive control of copper sulfate ($CuSO_4$) was used in a few trials at the observed Cu values of 65, 140, 335, and 885 mg Cu per 1 kg of soil. For the other tests, the maximum concentration was used [41]. Various studies showed different effects of CuNPs on various plants. It is listed in the given Table 1.

8 Conclusion and Future Perspectives

Copper nanoparticles (Cu NPs) have shown tremendous potential in agricultural applications due to their multifunctional qualities and unique properties. However, their extensive use raises concerns about their impact on soil microbial communities and overall soil health. The interactions of Cu NPs with soil constituents, including pH, soil type, organic matter content, contaminants, and environmental conditions, are crucial determinants of their solubility, bioavailability, and biotransformation. These factors significantly influence the nanoparticles' behavior in the soil–plant

Fig. 2 Effect of copper nanoparticles in agriculture increase plant productivity and growth stimulation

system, including their absorption, translocation, and distribution under various conditions. Furthermore, the antimicrobial properties of Cu NPs, while advantageous for pest and pathogen control, may inadvertently inhibit essential microbial activities, leading to long-term implications for soil fertility and ecosystem sustainability. Understanding the effects and mechanisms of Cu NPs in the soil–plant ecosystem is vital for developing strategies to mitigate potential risks while harnessing their benefits.

Table 1 Effects of CuO nanoparicles on plants

Effect of nanoparticles	Concentration and size of CuO nanoparticles	Plant	Key reference
Decrease in amino acid and ascorbic acid content	< 100 nm	*Pistia stratiotes L*	[29]
Decrease in root length, delayed senescence	10–1000 mg/L	Ipt-cotton	[22]
Reduction in shoot elongation and germination	Up to 80 mg/kg	*Coriandrum sativum*	[51]
Significant reduction in weight, root length, and chlorophyll content	Up to 500 mg/L	*G.max*	[27]
Inhibition of development, growth. Decrease in nutrition	Up to1000 mg/L	Transgenic and conventional cotton	[47]
Root and shoot growth affected	Up to 500 mg/L	*Brassica juncea*	[28]
Biomass, root length, chlorophyll content affected	200–500 mg/L	*Vigna radiata*	[36]
Elongation of root and seed germination inhibited	100–500 mg/L	*Cucumis sativus*	[26]
DNA damage induced, DNA lesion, accumulation	10–1000 mg/L	*L. rigidum, perenne, R. sativus*	[36]
Irregular shape of mitochondria, number of chlorophyll reduced, effect on chloroplast membrane	Up to 10 g/kg	*Landoltia punctata*	[21]
Increase in S, P, and Cu level in shoot, root length affected	Up to 20 mg/L	*M. sativa, Lactuca sativa*	[36]

Competing Interests The authors declare that they have no competing interest.

Declaration of Generative AI and AI-Assisted Technologies in the Writing Process During the preparation of this work the book chapter author(s) have not taken any assistance from AI, and the book chapter author(s) reviewed and edited the content as needed and take(s) full responsibility for the content of the publication. All authors read and approved the final version of the manuscript.

References

1. Argüello JM, Raimunda D, Padilla-Benavides T (2013) Mechanisms of copper homeostasis in bacteria. Front Cell Infect Microbiol 3:73
2. Bachheti RK, Bachheti A, Husen A (Eds.) (2023) In: Nanomaterials for environmental and agricultural sectors. Springer, Singapore. https://doi.org/10.1007/978-981-99-2874-3

3. Bachheti AJ, Bachheti RK, Husen A (Eds.) (2024) In: Carbon-based nanomaterials: synthesis, agricultural, biomedical, and environmental interventions. Springer Nature, Singapore. https://doi.org/10.1007/978-981-97-0240-4
4. Bachheti RK, Bachheti A, Husen A (Eds.) (2024a) In: Metal and metal-oxide based nanomaterials: synthesis, agricultural, biomedical and environmental interventions. Springer Nature, Singapore. https://doi.org/10.1007/978-981-99-7673-7
5. Bakshi M, Kumar A (2021) Copper-based nanoparticles in the soil-plant environment: assessing their applications, interactions, fate and toxicity. Chemosphere 281:130940
6. Bakshi M, Liné C, Bedolla DE (2019) Assessing the impacts of sewage sludge amendment containing nano-TiO_2 on tomato plants: a life cycle study. J Hazard Mater 369:191–198
7. Chandra S, Kumar A, Tomar PK (2014) Synthesis and characterization of copper nanoparticles by reducing agent. J Saudi Chem Soc 18:149–153
8. Cornelis G, Hund-Rinke K, Kuhlbusch T (2014) Fate and bioavailability of engineered nanoparticles in soils: a review. Crit Rev Environ Sci Technol 44:2720–2764
9. Fierer N, Jackson RB (2006) The diversity and biogeography of soil bacterial communities. Proc Natl Acad Sci USA 103:626–631
10. Frenk S, Ben-Moshe T, Dror I (2013) Effect of metal oxide nanoparticles on microbial community structure and function in two different soil types. PLoS ONE 8:e84441
11. Husen A, Siddiqi KS (2014) Phytosynthesis of nanoparticles: concept, controversy and application. Nanoscale Res Lett 9:229
12. Azamal H, Bachheti RK, Bachheti A (Eds.) (2023) In: Current trends in green nano-emulsions: food, agriculture and biomedical sectors. Singapore, Springer Nature. https://doi.org/10.1007/978-981-99-5398-1
13. Ingle AP, Duran N, Rai M (2014) Bioactivity, mechanism of action, and cytotoxicity of copper-based nanoparticles: a review. Appl Microbiol Biotechnol 98:1001–1009
14. Joshi A, Sharma A, Bachheti RK, Husen A, Mishra VK (2019) Plant-mediated synthesis of copper oxide nanoparticles and their biological applications. In: Nanomaterials and plant potential, Springer, Cham, pp 221–237. https://doi.org/10.1007/978-3-030-05569-1_8
15. Jośko I, Oleszczuk P, Futa B (2014) The effect of inorganic nanoparticles (ZnO, Cr_2O_3, CuO and Ni) and their bulk counterparts on enzyme activities in different soils. Geoderma 232–234:528–537
16. Kah M (2015) Nanopesticides and nanofertilizers: emerging contaminants or opportunities for risk mitigation? Front Chem 3:64
17. Kanhed P, Birla S, Gaikwad S (2014) In vitro antifungal efficacy of copper nanoparticles against selected crop pathogenic fungi. Mater Lett 115:13–17
18. Kasana RC, Panwar NR, Kaul RK, Kumar P (2017) Biosynthesis and effects of copper nanoparticles on plants. Environ ChemLett 15:233–240
19. Keller AA, Adeleye AS, Conway JR (2017) Comparative environmental fate and toxicity of copper nanomaterials. NanoImpact 7:28–40
20. Khorsand Zak A, Abd. Majid WH, Abrishami ME, Yousefi R (2011) X-ray analysis of ZnO nanoparticles by Williamson–Hall and size–strain plot methods. Solid State Sci 13:251–256
21. Lalau CM, Mohedano R de A, Schmidt ÉC (2015) Toxicological effects of copper oxide nanoparticles on the growth rate, photosynthetic pigment content, and cell morphology of the duckweed Landoltiapunctata. Protoplasma 252:221–229
22. Le Van N, Rui Y, Cao W (2016) Toxicity and bio-effects of CuO nanoparticles on transgenic Ipt-cotton. J Plant Interact 11:108–116
23. Lofts S, Criel P, Janssen CR (2013) Modelling the effects of copper on soil organisms and processes using the free ion approach: towards a multi-species toxicity model. Environ Pollut 178:244–253
24. Leyu AM, Debebe SE, Bachheti A, Rawat YS, Bachheti RK (2023) Green Synthesis of gold and silver nanoparticles using invasive alien plant *Parthenium hysterophorus* and their antimicrobial and antioxidant activities. Sustainability 15(12):9456
25. Miao A-J, Schwehr KA, Xu C (2009) The algal toxicity of silver engineered nanoparticles and detoxification by exopolymeric substances. Environ Pollut 157:3034–3041

26. Moon Y-S, Park E-S, Kim T-O (2014) SELDI-TOF MS-based discovery of a biomarker in Cucumissativus seeds exposed to CuO nanoparticles. Environ Toxicol Pharmacol 38:922–931
27. Nair PMG, Chung IM (2014) A mechanistic study on the toxic effect of copper oxide nanoparticles in soybean (Glycine max L.) root development and lignification of root cells. Biol Trace Elem Res 162:342–352
28. Nair PMG, Chung IM (2015) Study on the correlation between copper oxide nanoparticles induced growth suppression and enhanced lignification in Indian mustard (Brassica juncea L.). Ecotoxicol Environ Saf 113:302–313
29. Olkhovych O, Volkogon M, Taran N (2016) The effect of copper and zinc nanoparticles on the growth parameters, contents of ascorbic acid, and qualitative composition of amino acids and acylcarnitines in Pistiastratiotes L. (Araceae). Nanoscale Res Lett 11:218
30. Panyala NR, Peña-Méndez EM, Havel J (2008) Silver or silver nanoparticles: a hazardous threat to the environment and human health? J Appl Biomed 6:117–129
31. Parada J, Rubilar O, Diez MC (2019) Combined pollution of copper nanoparticles and atrazine in soil: effects on dissipation of the pesticide and on microbiological community profiles. J Hazard Mater 361:228–236
32. Pariona N, Mtz-Enriquez AI, Sánchez-Rangel D (2019) Green-synthesized copper nanoparticles as a potential antifungal against plant pathogens. RSC Adv 9:18835–18843
33. Peng C, Xu C, Liu Q (2017) Fate and transformation of CuO nanoparticles in the soil-rice system during the life cycle of rice plants. Environ Sci Technol 51:4907–4917
34. Pradhan A, Seena S, Pascoal C, Cássio F (2011) Can metal nanoparticles be a threat to microbial decomposers of plant litter in streams? MicrobEcol 62:58–68
35. Qu H, Ma C, Xing W (2022) Effects of copper oxide nanoparticles on Salix growth, soil enzyme activity and microbial community composition in a wetland mesocosm. J Hazard Mater 424:127676
36. Rajput VD, Minkina T, Sushkova S (2018) Effect of nanoparticles on crops and soil microbial communities. J Soils Sediments 18:2179–2187
37. Ren G, Hu D, Cheng EWC (2009) Characterisation of copper oxide nanoparticles for antimicrobial applications. Int J Antimicrob Agents 33:587–590
38. Rippner DA, Margenot AJ, Fakra SC (2021) Microbial response to copper oxide nanoparticles in soils is controlled by land use rather than copper fate. Environ Sci Nano 8:3560–3576
39. Rubina MS, Vasil'kov AY, Naumkin AV (2017) Synthesis and characterization of chitosan–copper nanocomposites and their fungicidal activity against two sclerotia-forming plant pathogenic fungi. J Nanostruct Chem 7:249–258
40. Saharan V, Sharma G, Yadav M (2015) Synthesis and in vitro antifungal efficacy of Cu–chitosan nanoparticles against pathogenic fungi of tomato. Int J BiolMacromol 75:346–353
41. Samarajeewa AD, Velicogna JR, Schwertfeger DM (2021) Ecotoxicological effects of copper oxide nanoparticles (nCuO) on the soil microbial community in a biosolids-amended soil. Sci Total Environ 763:143037
42. Shi J, Ye J, Fang H, (2018) Effects of copper oxide nanoparticles on paddy soil properties and components. Nanomaterials (Basel) 8. https://doi.org/10.3390/nano8100839
43. Simonin M, Cantarel AAM, Crouzet A (2018) Negative effects of copper oxide nanoparticles on carbon and nitrogen cycle microbial activities in contrasting agricultural soils and in presence of plants. Front Microbiol 9:3102
44. Simonin M, Richaume A (2015) Impact of engineered nanoparticles on the activity, abundance, and diversity of soil microbial communities: a review. Environ Sci Pollut Res Int 22:13710–13723
45. Tourinho PS, van Gestel CAM, Lofts S (2012) Metal-based nanoparticles in soil: fate, behavior, and effects on soil invertebrates. Environ ToxicolChem 31:1679–1692
46. Tripathi S, Champagne D, Tufenkji N (2012) Transport behavior of selected nanoparticles with different surface coatings in granular porous media coated with Pseudomonas aeruginosa biofilm. Environ Sci Technol 46:6942–6949
47. Van NL, Ma C, Shang J (2016) Effects of CuO nanoparticles on insecticidal activity and phytotoxicity in conventional and transgenic cotton. Chemosphere 144:661–670

48. Worku LA, Tadesse MG, Bachheti A, Pandey DP, Chandel AK, Ewuntu AW, Bachheti RK (2024) Experimental investigations on PVA/chitosan and PVA/chitin films for active food packaging using *Oxytenanthera abyssinica* lignin nanoparticles and its UV-shielding, antimicrobial, and antiradical effects. Int J Biol Macromol 254:127644. https://doi.org/10.1016/j.ijbiomac.2023.127644
49. Wassie AT, Ahmed IN, Bachheti A, Husen A, Bachheti RK (2024) Plant extract-assisted green synthesis and characterization of copper nanocomposite. In: Metal and metal-oxide based nanomaterials: synthesis, agricultural, biomedical and environmental interventions. Singapore, Springer Nature, pp 43–66. https://doi.org/10.1007/978-981-99-7673-7_3
50. Xu C, Peng C, Sun L (2015) Distinctive effects of TiO_2 and CuO nanoparticles on soil microbes and their community structures in flooded paddy soil. Soil BiolBiochem 86:24–33
51. Zuverza-Mena N, Medina-Velo IA, Barrios AC (2015) Copper nanoparticles/compounds impact agronomic and physiological parameters in cilantro (Coriandrumsativum). Environ Sci Process Impacts 17:1783–1793

Copper-Based Nanomaterials for Precision Agriculture and Smart Farming

Jije Mideksa Geyesa

Abstract Copper nanomaterials have received considerable attention in recent years due to their potential applications in precision agriculture and smart farming. A variety of applications of nanomaterials are discussed in this chapter, including crop protection, disease management, nutrient management, soil remediation, sensing and monitoring, and controlled release systems. Moreover, these materials can be applied to sensing and monitoring systems and can detect plant health conditions in real time. They are also used in the controlled, targeted release of agrochemicals, fertilizers, and growth regulators, resulting in improved agricultural productivity. The exploration of copper nanomaterials in precision agriculture and smart farming indicates potential for improving practices while minimizing environmental impact and ensuring sustainable farming. However, there are several concerns regarding using copper nanomaterial in agriculture, including toxicity and accumulation in soil. These problems can be resolved by encapsulating copper nanomaterials with biodegradable polymers and integrating technologies like nanorobotics and autonomous systems, which can promote minimized release and reduce unwanted effects on the plant and the environment. The aim of this chapter is to provide overview of copper-based nanomaterials for precision agriculture and smart farming.

Keywords Copper nanomaterial · Nutrient management · Controlled release systems · Soil remediation

J. M. Geyesa (✉)
Department of Industrial Chemistry, College of Applied Science, Addis Ababa Science and Technology University, Addis Ababa, Ethiopia
e-mail: jmgeyesa@mtu.edu; jijemideksa@gmail.com

Departement of Chemistry, Michigan Technological University, Houghton, USA

R. K. Bachheti et al. (eds.), *Copper-Based Nanomaterial for Agricultural Practices*, Smart Nanomaterials Technology, https://doi.org/10.1007/978-981-96-8610-0_13

1 Introduction

Copper nanomaterials can be used in soil health and nutrient management. Their unique features make them compatible for improving soil fertility, increasing plant consumption of nutrients, lessening nutrient losses, preventing environmental pollution and so on [16, 29, 37]. They also boost the physical, chemical, and biological aspects of soil by increasing nutrient availability and soil structure. Copper nanomaterials also increase nutrient absorption when applied directly to the soil, resulting in in better nutrient uptake by plant roots. This phenomenon indicates an improvement in agricultural yield and productivity due to these materials [18].

In addition, copper nanomaterials can also help reduce nutrient losses. Nutrient leaching and runoff are major concerns in agriculture. These occurrences cause environmental contamination and water pollution, and copper nanomaterials were found to decrease nutrient leaching by holding nutrients in the soil and preventing their movement into groundwater. This aids in guarding the surrounding ecosystems from contamination [35]. Another significant application of copper nanomaterial includes the prevention of environmental contamination. Commonly used fertilizers often release excess nutrients into the soil, causing imbalance and having a negative impact on the environment. These materials act as controlled systems by gradually releasing nutrients to the plants over an extended period. This mechanism minimizes nutrient losses, guarantees nutrient availability to the plants when needed, and reduces the risk of environmental contamination [30].

Furthermore, copper nanomaterials hold unique properties that enable their controlled release of copper ions. This mechanism helps maintain a steady and prolonged supply of copper ions to plants, which is toxic to pathogens and pests. This results in less risk of phytotoxicity to the plants.

2 Nanomaterials for Precision and Smart Farming

Nanomaterials have magnetic, electric, optical, mechanical, and catalytic residences, which differ from their bulk opposite numbers. These materials have at least one dimension starting from 1 to a hundred nm. This makes them flexible to manipulate the houses of nanomaterials to attain the preferred consequences [5].

Precision farming refers to the right use of agricultural inputs depending on soil, weather, and crop wishes so one can make the most of lengthy-time period crop output, high quality, and profitability. It permits farmers to gather the blessings of modern technologies while keeping the first-rate of their land and produce [25]. Nanomaterials have additionally shown amazing potential in precision and clever farming practices. They provide unique and high-quality houses, which could drastically enhance diverse agricultural factors. Applications of nanomaterials for precision and clever framing practices are proven in Fig. 1

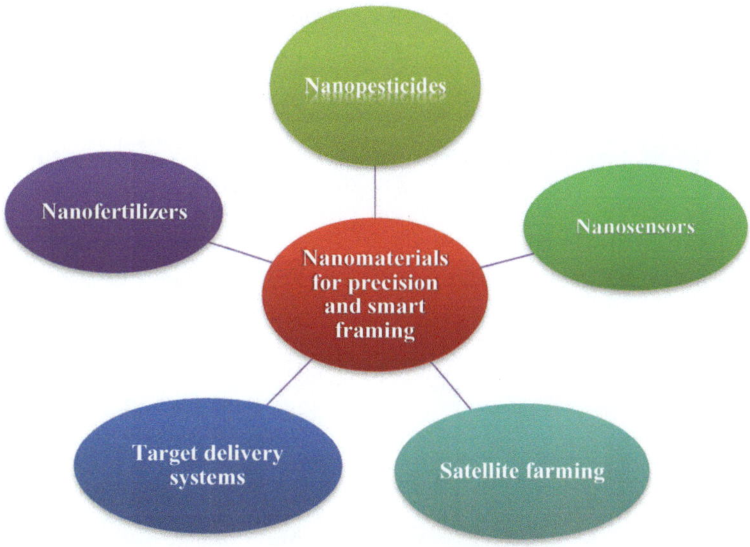

Fig. 1 Application of nanomaterials for precision and smart framing practices

2.1 Nanopesticides

Nanotechnology can increase the effectiveness of pesticides while minimizing their bad effect on the environment. Nanopesticides involve the formulation of pesticide particles at the nanoscale. These nanomaterials can enhance the amount of insecticides required to be less than traditional pesticides [8]. The Applications of nanopesticide are illustrated in (Fig. 2).

2.2 Nanofertilizers

Nanofertilizers are designed to boost plant nutrient availability and uptake. This can be performed by encapsulating nutrients within nanomaterials. Their controlled release can be attained by allowing for the sustained and targeted supply of nutrients to plants. This way of nutrient delivery can help improve plant growth [10]. Benefits of using nanofertilizers are depicted in (Fig. 3).

2.3 Nanosensors

Nanosensors are a class of nanomaterials that can provide real-time data on growing conditions and plant parameters. These sensors can be engrained in the soil, leaves,

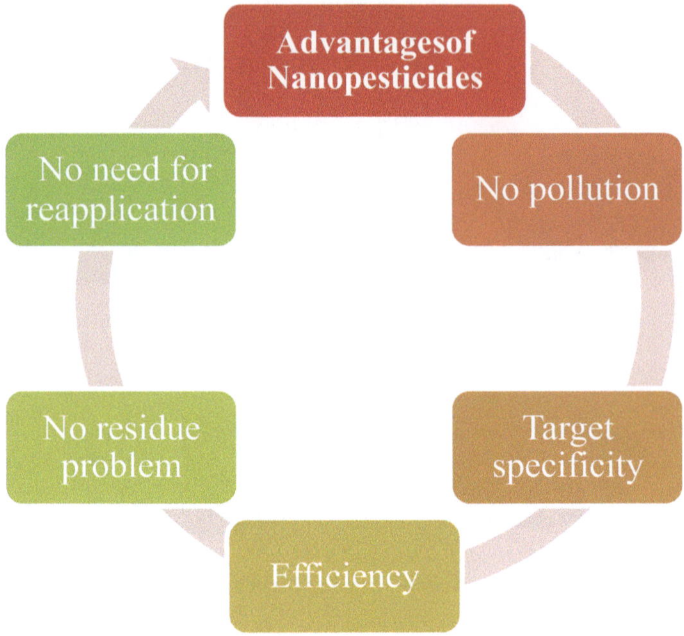

Fig. 2 Adantages of nanopesticides

Fig. 3 Advantages of nanofertilizers

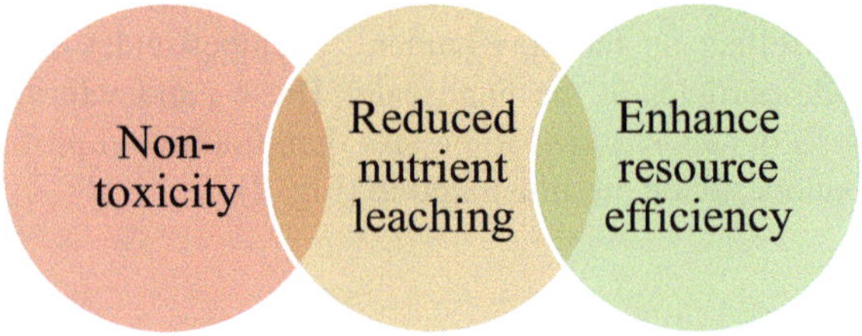

and plants, measuring moisture levels, temperature, nutrient concentration, and pest presence.

The data gathered allows farmers to make informed decisions about irrigation, fertilization, and crop protection. This is helpful in optimizing resource use, minimizing waste, and improving crop yields [17].

2.4 *Nanomaterial Enabled Target Delivery*

Nanoparticles can encapsulate, which helps them reach their proposed target in a controlled and targeted manner. For instance, nanocapsules can deliver pesticides directly to pest-infested areas, reducing overall pesticide use. Targeted release aids in minimizing environmental pollution and supports nutrient management strategies [27].

2.5 *Satellite Farming*

Satellite farming is another name for precision agriculture, and it includes the use of satellite imagery and remote sensing technologies to manage agricultural practices. Satellites provide high-resolution images that help detect vegetation health, analyze nutrient deficiencies, forecast crop yields, and determine areas of disease spread. This farming type helps farmers make decisions based on resource allocation, water, fertilizer, and pathogen control [31]. The benefits of satellite farming are depicted in (Fig. 4).

3 Copper Nanomaterials

Copper nanomaterials are particles formed of copper with nanoscale dimensions (fewer than 100 nm). Their small size and large surface area differentiate them from bulk copper [31].They have several classifications, each with unique physical and chemical features that result in their usefulness in a variety of applications [28] (Fig. 5 depicts this.)

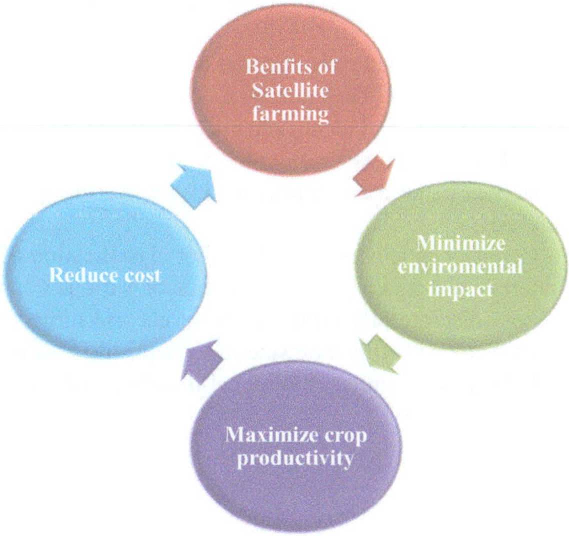

Fig. 4 Benefits of satellite farming

Fig. 5 Copper nanomaterial classification

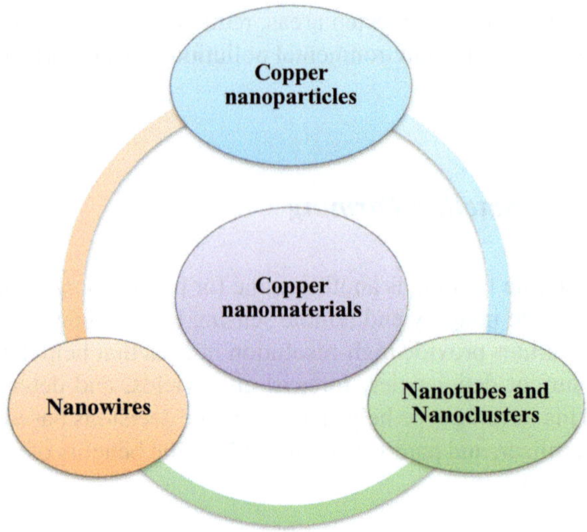

Copper nanomaterials are commonly used materials. Various methods can be used to make copper nanomaterials, including Chemical reduction, electrochemical deposition, and thermal evaporation. These materials have strong electrical and thermal conductivity and antibacterial properties. They are utilized in various applications, including electronics, catalysis, energy storage, and biomedical disciplines [22].

Copper nanowires are among copper nanomaterials. They are characterized by their elongated structures with nanoscale dimensions and lengths of several micrometers. They have high electrical conductivity and flexibility. These materials are used in the production of transparent conductive films, touchscreens, and flexible electronics [36].

Another copper nanomaterial is copper nanotubes, which are tubular-shaped hollow structures. They can be produced by different methods, including template-based synthesis and electrochemical methods. They also have unique features such as high aspect ratios, strong electrical conductivity, and a huge surface area. These properties make them applicable in the making of sensors, batteries, and selective catalysis.

Copper nanoclusters are the other classification of copper nanomaterials. These materials are small, stable groupings of various copper atoms. They are known to possess discrete electrical energy levels and molecular-like features. They can be assembled by varying their size and composition. Applications of copper nanoclusters include sensing, optics, biological imaging, and many others [4].

4 Application of Copper-Based Nanomaterials for Precision Agriculture and Smart Farming

Copper-based nanomaterials have various applications in the agriculture industry. These include efficient nutrition and targeted pesticide delivery, risk mitigation, and plant growth. They play a significant role in overcoming unsolved agricultural problems. Figure 6 illustrates the use of copper-based nanomaterials in precision agriculture and smart farming.

4.1 Crop Protection and Disease Management

Copper-based nanomaterials are capable of replacing traditional insecticides. They have antibacterial and antifungal properties that help prevent and control plant fungal and bacterial infections. In addition, their nanoscale size enables efficient delivery and absorption by plant tissues. Copper-based compounds have long been used as fungicides and bactericides in agriculture. However, traditional copper formulations can cause environmental concerns since they can be stored in soil and water bodies, which cause hazardous effects on non-target plant parts. These materials have advantages in terms of controlled release and surface area-to-volume ratio. A modest amount of copper can be used to protect crops effectively without damaging non targeted plant parts. Furthermore, nanomaterials can target infections while reducing any unwanted side effects on beneficial organisms. They can also be designed to have

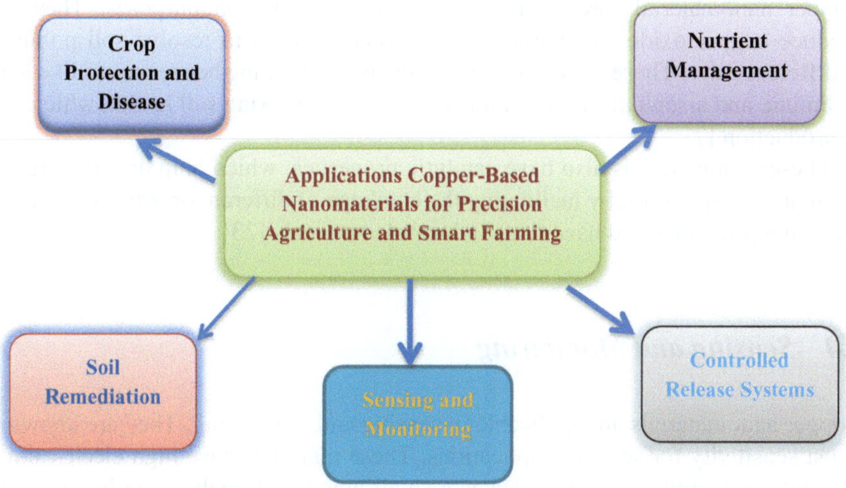

Fig. 6 Application of copper-based nanomaterials for precision agriculture and smart

upgraded stability and sustained release qualities, which will aid in more effective and longer-lasting anti-pathogen action [36].

Copper nanomaterials also have applications in soil management. Agricultural soils usually encounter nutrient imbalances and heavy metal contamination. When copper nanomaterials are incorporated, they help remedy heavy metal contamination by absorption and immobilization of these unwanted toxins. They can operate in this way because of their high surface area, catalytic activity, and antimicrobial properties [26].

4.2 Nutrient Management

Copper nanomaterials can be used to manage nutrient availability and absorption in agricultural systems. These materials increase nutrient interactions in soil by boosting nutrient delivery to plants, particularly for minerals that are easily leached in soil. They also enhance nutrient solubility and mobility, resulting in improved accessibility to plant roots [14].

Copper nanomaterials can also act as catalysts in nutritional processes. These materials aid in the conservation of nutrient forms to plant viable forms, which helps solve the deficiency problem that is common in agricultural soils [15].

4.3 Soil Remediation

Copper nanomaterials are applied to soil for remediation purposes. They can produce copper oxide nanoparticles, which can be used to resolve soil pollution in different ways. These materials are capable of eliminating heavy metals like lead, cadmium, and arsenic from contaminated soils and restoring soil health, which aids in production [7].

These nanomaterials also have catalytic properties, which help degrade organic pollutants found in soil by facilitating the breakage of different organic compounds, including pesticides and hazardous industrial chemicals [23].

4.4 Sensing and Monitoring

Copper nanomaterials are applicable in sensing and monitoring. They are known to boost sensitivity for sensing applications. These materials have high electrical and thermal conductivity, which indicates a small amount of analyte can be measured accurately when copper nanomaterials are used for sensing purposes [6].

Copper nanomaterials have also been used in biosensing applications. This is because their large surface area can enhance the immobilization of enzymes, antibodies, and other biological molecules. This is particularly useful in biomedical and environmental monitoring [19].

Copper nanomaterials also have electrochemical sensing properties. They can detect various analytes like heavy metals, organic chemicals, and biological indicators. They change the surface of copper nanomaterials with a specific functional group, which aids in accurate and reliable detection [37]. Copper nanowires and nanosensors can be incorporated into agricultural devices and systems for monitoring soil, water quality, and plant condition parameters [12].

4.5 Controlled Release Systems

Copper nanomaterials have attracted considerable interest as potential controlled-release systems for insecticides, herbicides, and agrochemicals. These materials have distinct advantages over traditional delivery methods, including increasing active chemicals' efficacy, lowering their environmental impact, and boosting overall agricultural sustainability [35].

Controlled-release systems can be formed by adsorbing or binding these compounds to the surface of nanomaterials. The high surface area-to-volume ratio of copper nanomaterials is an important feature. This characteristic makes it possible to efficiently load insecticides, herbicides, and other agrochemicals onto nanomaterials [10].

The controlled-release process of copper nanomaterials comprises the slow release of loaded agrochemicals over time. This controlled release can be performed via different methods, such as diffusion of active substances through pores and channels within the nanomaterials, surface erosion, and slow destruction of the copper nanomaterials themselves [3].

Furthermore, copper nanomaterials can lower the amount of agrochemicals required in agricultural methods. Compared to traditional application methods, the controlled-release system provides for lower dosage requirements. This reduction in chemical use can help reduce environmental pollution and the risk of non-target organisms being exposed to these chemicals [1].

5 Copper Nanomaterial Challenges and Prospects for Precision and Smart Farming

Copper nanomaterials have advantages and drawbacks. One of the most concerning problems with using these materials in agriculture is their accumulation in soil and water systems. This results in higher unwanted quantities of copper nanomaterial retention, which harms ecosystems [2].

To solve this problem, researchers are searching for various approaches to lessen the environmental risk associated with copper nanomaterials. They have tried to encapsulate copper nanomaterials in biodegradable polymer, which stabilizes the surface, making it less reactive and helping to reduce the release to the environment. Additionally, copper nanomaterials' regulated and targeted use can reduce total utilization and potential soil accumulation.

6 Conclusion

Copper nanomaterials have a significant role in transforming precision agriculture and smart farming techniques. Due to their size, shape, and reactivity, these materials have unique features, including strong reactivity, antibacterial activity, and catalytic capability, which make them applicable in various agricultural applications. Farmers' use of copper nanomaterials will increase plant development and aid in plant disease management, boosting overall agricultural production. In addition, Copper nanomaterials prevent the growth of bacteria, Fungi, and viruses, which makes them more advantageous than chemical pesticides. This aids in reducing the usage of toxic substances, which reduces possible environmental impact. Copper nanomaterial combined with nanosensors, robots, and data analytics can allow for real-time monitoring of plant health, nutrient levels, and environmental conditions.

Competing Interests The authors declare that they have no competing interest.

Declaration of Generative AI and AI-Assisted Technologies in the Writing Process During the preparation of this work the book chapter author(s) have not taken any assistance from AI, and the book chapter author(s) reviewed and edited the content as needed and take(s) full responsibility for the content of the publication. All authors read and approved the final version of the manuscript.

References

1. Abd-Elsalam KA (2022) Copper-based nanomaterials: next-generation agrochemicals: a note from the editor. In: Copper nanostructures: next-generation of agrochemicals for sustainable agro ecosystems. Elsevier, pp 1–14
2. Ali Q, Zia MA, Kamran M, Shabaan M, Zulfiqar U, Ahmad M, Maqsood MF (2023) Nanoremediation for heavy metal contamination: a review. Hybrid Adv 100091

3. Ashitha A, Mathew J (2020) Characteristics and types of slow/controlled release of pesticides. In: Controlled release of pesticides for sustainable agriculture, pp 141–153
4. Babu Busi K, Palanivel M, Kanta Ghosh K, Basu Ball W, Gulyás B, Padmanabhan P, Chakrabortty S (2022) The multifarious applications of copper nanoclusters in biosensing and bioimaging and their translational role in early disease detection. Nanomaterials 12(3):301
5. Baig N, Kammakakam I, Falath W (2021) Nanomaterials: a review of synthesis methods, properties, recent progress, and challenges. Mater Adv 2(6):1821–1871
6. Bakshi M, Kumar A (2022) Applications of copper nanoparticles in plant protection and pollution sensing: toward promoting sustainable agriculture. In: Copper nanostructures: next-generation of agrochemicals for sustainable agro ecosystems. Elsevier, pp 393–413
7. Banda M, Munyengabe A, Augustyn W (2023) Aromatic plants: alternatives for management of crop pathogens and ideal candidates for phytoremediation of contaminated land
8. Chaudhary A, Krishnarth N, Nath P (2023) 3Nanopesticides. In: Nanopesticides, nanoherbicides, and nanofertilizers: formulations and applications
9. Fincheira P, Hoffmann N, Tortella G, Ruiz A, Cornejo P, Diez MC, Rubilar O (2023) Eco-efficient systems based on nanocarriers for the controlled release of fertilizers and pesticides: toward smart agriculture. Nanomaterials 13(13):1978
10. Gomes DG, Pieretti JC, Lourenço IM, Oliveira HC, Seabra AB (2022) Copper-based nanoparticles for pesticide effects. Inorganic Nanopesticides and nanofertilizers: a view from the mechanisms of action to field applications. Springer International Publishing, Cham, pp 187–212
11. Gondal AH, Tayyiba L (2022) Prospects of using nanotechnology in agricultural growth, environment and industrial food products. Rev Agricul Sci 10:68–81
12. Hameed A, Saif MJ, Qayyum MA, Khalid T, Farooq T (2023) Nanomaterial-based sensors for real-time monitoring of crop plants growth, development, production, and protection. In: Engineered nanomaterials for sustainable agricultural production, soil improvement and stress management. Academic Press, pp 357–385
13. Humbal A, Pathak B (2023) Application of nanotechnology in plant growth and diseases management: tool for sustainable agriculture. In: Agricultural and environmental nanotechnology: novel technologies and their ecological impact. Singapore, Springer Nature, pp 145–168
14. Hyder S, Ul-Nisa M, Shahid H, Gohar F, Gondal AS, Riaz N, Iqbal R (2023) Recent trends and perspectives in the application of metal and metal oxide nanomaterials for sustainable agriculture. Plant Physiol Biochem 107960
15. Ingle AP (ed) (2021) Nanotechnology in plant growth promotion and protection. Wiley & Sons, Incorporated
16. Joshi A, Sharma A, Bachheti RK, Husen A, Mishra VK (2019) Plant-mediated synthesis of copper oxide nanoparticles and their biological applications. In: Nanomaterials and plant potential, Springer, Cham, pp 221–237. https://doi.org/10.1007/978-3-030-05569-1_8
17. Johnson MS, Sajeev S, Nair RS (2021) Role of nanosensors in agriculture. In: 2021 International conference on computational intelligence and knowledge economy (ICCIKE), March, IEEE, pp 58–63
18. Kopittke PM, Lombi E, Wang P, Schjoerring JK, Husted S (2019) Nanomaterials as fertilizers for improving plant mineral nutrition and environmental outcomes. Environ Sci Nano 6(12):3513–3524
19. Kumar V, Raghuwanshi SK, Kumar S (2022) Recent advances in carbon nanomaterials based spr sensor for biomolecules and gas detection-A review. IEEE Sensors J
20. Kumari B, Prakash A (2023) Sustainable nanotechnology in agriculture. In: Nanopesticides, nanoherbicides, and nanofertilizers. CRC Press, pp 26–37
21. Li Y, Zhang P, Li M, Shakoor N, Adeel M, Zhou P, Rui Y (2023) Application and mechanisms of metal-based nanoparticles in the control of bacterial and fungal crop diseases. Pest Manag Sci 79(1):21–36
22. Maliki M, Ifijen IH, Ikhuoria EU, Jonathan EM, Onaiwu GE, Archibong UD, Ighodaro A (2022) Copper nanoparticles and their oxides: optical, anticancer and antibacterial properties. Int Nano Lett 12(4):379–398

23. Mazarji M, Minkina T, Sushkova S, Mandzhieva S, Bidhendi GN, Barakhov A, Bhatnagar A (2021) Effect of nanomaterials on remediation of polycyclic aromatic hydrocarbons-contaminated soils: a review. J Environ Manage 284:112023
24. Mohamed HI, Sajyan TK, Shaalan R, Bejjani R, Sassine YN, Basit A (2022) Plant-mediated copper nanoparticles for agri-ecosystem applications. In: Agri-waste and microbes for production of sustainable nanomaterials.Elsevier, pp 79–120
25. Raj, E. F. I., Appadurai, M., & Athiappan, K. (2022). Precision farming in modern agriculture. In Smart Agriculture Automation Using Advanced Technologies: Data Analytics and Machine Learning, Cloud Architecture, Automation and IoT (pp. 61–87). Singapore: Springer Singapore.
26. Rippner DA, Margenot AJ, Fakra SC, Aguilera LA, Li C, Sohng J, Parikh SJ (2021) Microbial response to copper oxide nanoparticles in soils is controlled by land use rather than copper fate. Environ Sci Nano 8(12):3560–3576
27. Saleh MM, Mahmoud AS, Abbas HS, Abu-Ellail FF, Kotakonda M, Salem KF (2022) Nanotechnological approaches for efficient delivery of plant ingredients. Sustainable agriculture reviews 53: nanoparticles: a new tool to enhance stress tolerance. Springer International Publishing, Cham, pp 247–286
28. Shaheen I, Khalil A, Shaheen R, Tahir MB (2023) A review on nanomaterials: types, synthesis, characterization techniques, properties and applications. Innov Sci Technol 2(1):56–62
29. Siddiqi KS, Husen A (2020) Current status of plant metabolite-based fabrication of copper/copper oxide nanoparticles and their applications: a review. Biomater Res 24(11):1–15. https://doi.org/10.1186/s40824-020-00188-1
30. Singh P, Singh AP (2021) Nanomaterials in soil health management and crop production: potentials and limitations. In: Handbook of nanomaterials and nanocomposites for energy and environmental applications, pp 1221–1245
31. Sishodia RP, Ray RL, Singh SK (2020) Applications of remote sensing in precision agriculture: a review. Remote Sensing 12(19):3136
32. Arkoti NK, Verma V, Pal K (2022) Nanomaterials and their distinguishing features. In: Nanomaterials for advanced technologies. Singapore, Springer Nature, pp 1–18; Precision agriculture: a review. Remote Sens 12(19):3136
33. Strayer-Scherer A, Timilsina S, Liao YY, Young M, Rosskopf EN, Vallad GE, Paret ML (2022) Simulated leaching of foliar applied copper bactericides on the soil micro biome utilizing various beta diversity resemblance measurements. Microbiol Spectrum 10(3):e01481-e1521
34. Tufail MA, Iltaf J, Zaheer T, Tariq L, Amir MB, Fatima R, Ayyub M (2022) Recent advances in bioremediation of heavy metals and persistent organic pollutants: a review. Sci Total Environ 850:157961
35. Ur Rahim H, Qaswar M, Uddin M, Giannini C, Herrera ML, Rea G (2021) Nano-enable materials promoting sustainability and resilience in modern agriculture. Nanomaterials 11(8):2068
36. Wang J, Piao W, Jin X, Jin LY, Yin Z (2022) Recent progress in metal nanowires for flexible energy storage devices. Front Chem 10:920430
37. Wassie AT, Ahmed IN, Bachheti A, Husen A, Bachheti RK (2024) Plant extract-assisted green synthesis and characterization of copper nanocomposite. In: Metal and metal-oxide based nanomaterials: synthesis, agricultural, biomedical and environmental interventions. Singapore, Springer Nature, pp 43–66. https://doi.org/10.1007/978-981-99-7673-7_3

MIX
Papier aus verantwortungsvollen Quellen
Paper from responsible sources
FSC® C105338

If you have any concerns about our products,
you can contact us on
ProductSafety@springernature.com

In case Publisher is established outside the EU,
the EU authorized representative is:
Springer Nature Customer Service Center GmbH
Europaplatz 3, 69115 Heidelberg, Germany

Printed by Libri Plureos GmbH
in Hamburg, Germany